Science Framework for California Public Schools

Kindergarten Through Grade Twelve

With New Criteria for Instructional Materials

Developed by the
Curriculum Development and Supplemental Materials Commission

Adopted by the
California State Board of Education

Published by the
California Department of Education

Publishing Information

On March 10, 2004, the California State Board of Education approved the modified *Criteria for Evaluating Instructional Materials in Science, Kindergarten Through Grade Eight,* The following persons were serving on the State Board at that time: Reed Hastings, President; Joe Nuñez, Vice President; Ruth Bloom; Don Fisher; Ruth E. Green; Glee Johnson; Jeannine Martineau; Bonnie Reiss; Suzanne Tacheny; Johnathan Williams. The new criteria are included in this edition of the framework.

When the *Science Framework for California Public Schools: Kindergarten Through Grade Twelve* was adopted by the California State Board of Education on February 6, 2002, the members of the State Board were the following: Reed Hastings, President; Joe Nuñez, Vice-President; Robert Abernethy; Donald G. Fisher; Susan Hammer; Nancy Ichinaga; Carlton Jenkins; Marion Joseph; Vicki Reynolds; Suzanne Tacheny; and Erika Goncalves.

The framework was developed by the Curriculum Development and Supplemental Materials Commission. (See pages vi–viii for the names of the members of the commission and the names of the principal writers and others who made significant contributions to the framework.)

This publication was edited by Faye Ong and Janet Lundin, working in cooperation with Greg Geeting, Assistant Executive Director, State Board of Education; Tom Adams, Administrator, and Christopher Dowell, Education Programs Consultant, Curriculum Frameworks and Instructional Resources Division, California Department of Education. It was designed and prepared for printing by the staff of CDE Press, with the cover and interior design created and prepared by Paul Lee. Typesetting was done by Jeannette Huff and Carey Johnson. The framework was published by the Department of Education, 1430 N Street, Sacramento, California 95814-5901. It was distributed under the provisions of the Library Distribution Act and *Government Code* Section 11096.

© 2004 by the California Department of Education
All rights reserved

ISBN 978-0-8011-1599-8

Ordering Information

Copies of this publication are available for $17.50 each, plus shipping and handling charges. California residents are charged sales tax. Orders may be sent to the California Department of Education, CDE Press, Sales Office, 1430 N Street, Suite 3207, Sacramento, CA 95814-5901; FAX (916) 323-0823.

An illustrated *Educational Resources* catalog describing publications, videos, and other instructional media available from the Department can be obtained without charge by writing to the address given above or by calling the Sales Office at (916) 445-1260.

Photo Credits

The California Department of Education gratefully acknowledges the National Oceanic and Atmospheric Administration for the use of the photographs on pages 34, 63, 64, 70, 73, 74, 89, 234, 259, 266, and 270; the National Aeronautics and Space Administration for the use of photographs on pages 52, 77, 81 (the background), 94, 137, 172, 251, 255, 274, and 299 (the background); and Paul A. Thiessen <http://www.ChemicalGraphics.com> for the use of the illustrations on pages 153 and 224.

Prepared for publication by CSEA members

Contents

Foreword v
Acknowledgments vi
State Board of Education Policy on the Teaching of Natural Sciences ix

1 Introduction to the Framework 1
Audiences for the Framework 2
Instructional Materials 3
The Challenges in Science Education 3
Science and the Environment 8
Guiding Principles 9
Organization of the Framework 13

2 The Nature of Science and Technology 15
The Scientific Method 16
Scientific Practice and Ethics 18
Science and Technology 18
Resources for Teaching Science and Technology 20
Science and Society 20

3 The Science Content Standards for Kindergarten Through Grade Five 23
Kindergarten 26
Grade One 31
Grade Two 36
Grade Three 45
Grade Four 56
Grade Five 66

4 The Science Content Standards for Grades Six Through Eight 81
Grade Six: Focus on Earth Sciences 84
Grade Seven: Focus on Life Sciences 103
Grade Eight: Focus on Physical Sciences 125

5 The Science Content Standards for Grades Nine Through Twelve 153
Physics 156
Chemistry 185
Biology/Life Sciences 220
Earth Sciences 251
Investigation and Experimentation 278

6 Assessment of Student Learning 281
Purposes of Assessment 282
Science Assessment Strategies 284
STAR Program Results 285
Summary 285

7 Universal Access to the Science Curriculum 287
Science and Basic Skills Development 288
Academic Language Development 288
English Learners 289
Advanced Learners 289
Students with Disabilities 289

8 Professional Development 293
What Is Professional Development? 295
Who Should Teach the Teachers? 296
When Is a Program Aligned with the Science Content Standards? 296
When Is a Professional Development Program Deemed Successful? 297
How Will Tomorrow's Science Teachers Be Developed? 297

9 Criteria for Evaluating Instructional Materials in Science, Kindergarten Through Grade Eight 299
Criteria Category 1: Science Content/Alignment with Standards 302
Criteria Category 2: Program Organization 303
Criteria Category 3: Assessment 304
Criteria Category 4: Universal Access 304
Criteria Category 5: Instructional Planning and Support 306

Selected References 309

Appendix
Requirements of the *Education Code* 310

Foreword

"The education of young people in science is at least as important, maybe more so, than the research itself."

—Glenn T. Seaborg

In 1998, with the adoption of the *Science Content Standards for California Public Schools*, California made a commitment that we will provide all students a world-class science education. Now, more than ever before, our students need a high degree of science literacy. This updated edition of the *Science Framework for California Public Schools* builds on the content standards, provides guidance for the education community to achieve that objective, and includes the 2004 evaluation criteria for the kindergarten-through-grade-eight instructional materials adopted by the State Board of Education. This framework is California's blueprint for our science curriculum, instruction, professional preparation and development, and instructional materials.

The framework offers guidance for science instruction in elementary, middle, and high schools. In kindergarten through grade five, students receive a solid foundation and acquire basic knowledge regarding physical, life, and earth sciences as well as learn investigation and experimentation skills.

Science instruction increases in depth and complexity in the middle grades, where the emphasis is on one science strand each year. In grade six, students focus on earth sciences; in grade seven, on life sciences; and in grade eight, on physical sciences. The investigation and experimentation standards increase in sophistication in the middle grades and require students to formulate a hypothesis for the first time, communicate the logical connections among hypotheses, and apply their knowledge of mathematics to analyze and report on data from their experiments.

At the high school level, science content is presented as four separate strands—physics, chemistry, biology/life sciences, and earth sciences—each providing students the rigor they need to prepare for collegiate-level study and career pathways. Both the content standards and the framework are designed so that the standards can be organized either as strand-specific courses or as courses that draw content from several strands. The high school investigation and experimentation standards ensure that students have experience in a laboratory setting.

The framework addresses a number of audiences—teachers, administrators, instructional materials developers, professional development providers, parents, guardians, and students. It makes the important point that science education must take place in conjunction with other core subjects, not in isolation from them.

This document establishes guiding principles that define the attributes of a quality science curriculum at all grade levels. The framework reflects the fundamental belief that all students can acquire the science knowledge and skills needed to succeed in the world that awaits them.

JACK O'CONNELL
State Superintendent of Public Instruction

RUTH E. GREEN
President, California State Board of Education

Acknowledgments

When the *Criteria for Evaluating Instructional Materials in Science, Kindergarten Through Grade Eight* was adopted by the California State Board of Education on March 10, 2004, the following persons were serving on the State Board:

Reed Hastings, President
Joe Nuñez, Vice President
Ruth Bloom, Member
Don Fisher, Member
Ruth E. Green, Member
Glee Johnson, Member
Jeannine Martineau, Member
Bonnie Reiss, Member
Suzanne Tacheny, Member
Johnathan Williams, Member

The new criteria are included in this version of the framework.

Members of the Curriculum Development and Supplemental Materials Commission (Curriculum Commission) serving at the time the criteria were recommended for approval to the State Board were:

Edith Crawford, San Juan Unified School District, *Chair*
Norma Baker, Los Angeles Unified School District, *Vice Chair*
William D. Brakemeyer, Fontana Unified School District
Milissa Glen-Lambert, Los Angeles Unified School District
Kerry Hamill, Oakland Unified School District,
Deborah Keys, Oakland Unified School District
Wendy Levine, Inglewood Unified School District
Sandra Mann, San Diego City Unified School District
Julie Maravilla, Los Angeles Unified School District
Michael Matsuda, Anaheim Union High School District
Mary-Alicia McRae, Salinas City Elementary School District
Stan Metzenberg, California State University, Northridge
Charles Munger, Stanford Linear Accelerator Center
Rosa Perez, Canada College
Joseph (Jose) Velasquez, Los Angeles Unified School District
Richard Wagoner, Los Angeles Unified School District

Members of the Science Subject Matter Committee of the Curriculum Commission responsible for overseeing the development of the *Criteria for Evaluating Instructional Materials in Science, Kindergarten Through Grade Twelve* were:

Sandra Mann, San Diego City Unified School District, *Chair*
Charles Munger, Stanford Linear Accelerator Center, *Vice Chair*
Michael Matsuda, Anaheim Union High School District
Stan Metzenberg, California State University, Northridge
Richard Wagoner, Los Angeles Unified School District

Special appreciation is extended to:

Rae Belisle, Executive Director, State Board of Education, for her leadership and collaboration with representatives from the state science community in revising the criteria
Sue Stickel, Deputy Superintendent, Curriculum and Instruction Branch, California Department of Education, for her assistance in revising the criteria

Note: The titles and affiliations of persons named in this list were current at the time the document was developed.

The development of the *Science Framework* involved the effort, dedication, and expertise of many individuals. The original draft of the framework was prepared between October 1998 and December 2000 by the Science Curriculum Framework and Criteria Committee (CFCC), a diverse group that included teachers and school administrators, university faculty, representatives of the business community, and parents (guardians) of students in public schools. The State Board of Education and the Curriculum Commission commend the following members of the Science CFCC and extend great appreciation to them:

Michael Rios, Montebello Unified School District, *Chair*

Richard Berry, San Diego State University, *Vice Chair*

Wallace Boggess, Simi Valley Unified School District

Robert Bornick, California State University, Northridge

Liselle Clark, Stockton Unified School District

Richard Feay, Los Angeles Unified School District

Jonathan Frank, San Francisco Unified School District

Hanna Hoffman, San Jose, California

Rita Hoots, Yuba College

G. Bradley Huff, Fresno County Office of Education

Linda Kenyon, Ceres Unified School District

Mary Kiely, Stanford University

Janet Kruse, Discovery Museum, Sacramento, California

Charles Munger, Stanford University

Eric Norman, Lawrence Berkeley National Laboratory

Keith Nuthall, Poway Unified School District

Lynda Rogers, San Lorenzo Unified School District

Ursula Sexton, San Ramon Unified School District

Steve Shapiro, Los Angeles, California

Jody Skidmore, North Cow Creek Elementary School District

William Tarr, Los Angeles Unified School District

Lane Therrell, Placerville, California

Commendation and appreciation are also extended to the principal writers of the *Science Framework:*

Roland (Rollie) Otto, Center for Science and Engineering Education, Lawrence Berkeley National Laboratory

Kathleen Scalise, Lawrence Hall of Science, University of California, Berkeley

Lynn Yarris, Lawrence Berkeley National Laboratory

The members of the Curriculum Commission's Science Subject Matter Committee (SMC), along with 2001 Curriculum Commission Chair **Patrice Abarca,** provided outstanding leadership in overseeing the development and editing of the *Science Framework:*

Richard Schwartz, Torrance Unified School District, *Science SMC Chair*

Veronica Norris, Tustin, California, *Science SMC Vice Chair*

Catherine Banker, Upland, California

Edith Crawford, San Juan Unified School District

Viken "Vik" Hovsepian, Glendale Unified School District

Other members of the Curriculum Commission who were serving at the time the *Science Framework* was recommended for approval to the State Board were:

Patrice Abarca, Los Angeles Unified School District, *2001 Chair*

Roy Anthony, Grossmont Union High School District

Marilyn Astore, Sacramento County Office of Education, *2000 Chair*

Rakesh Bhandari, Los Altos, California

Mary Coronado Calvario, Sacramento City Unified School District

Milissa Glen-Lambert, Los Angeles Unified School District

Lora L. Griffin, Retired Educator, Sacramento City Unified School District

Janet Philibosian, Los Angeles Unified School District

Leslie Schwarze, Trustee, Novato Unified School District

Susan Stickel, Elk Grove Unified School District, *2001 Vice Chair*

Karen S. Yamamoto, Washington Unified School District

Other individuals who served on the Curriculum Commission during the period of the *Science Framework's* development (1999-2000) were:

Eleanor Brown, San Juan Unified School District, *1999 Chair*

Ken Dotson, Turlock Joint Elementary School District

Lisa Jeffery, Los Angeles Unified School District

Joseph Nation, San Rafael, California

Barbara Smith, San Rafael City Elementary and High School District

Sheri Willebrand, Ventura Unified School District

The Curriculum Commission and its Science SMC received outstanding support and assistance from the State Board's curriculum liaison, **Marion Joseph.**

Final revision and editing of the *Science Framework* were conducted under the oversight of the State Board's science liaison, **Robert J. Abernethy,** who was assisted by **Greg Geeting,** Assistant Executive Director of the State Board, and **Charles Munger,** Stanford University.

During the review process, the Science SMC was aided by the following individuals who provided technical advice on science content issues:

Douglas E. Hammond, University of Southern California

William Hamner, University of California, Los Angeles

Eric Kelson, California State University, Northridge

George I. Matsumoto, Monterey Bay Aquarium Research Institute

Stan Metzenberg, California State University, Northridge

Martha Schwartz, University of Southern California

Steve Strand, University of California, Los Angeles

California Department of Education staff members who contributed to the *Science Framework's* development included:

Joanne Mendoza, Deputy Superintendent, Curriculum and Instructional Leadership Branch

Sherry Skelly Griffith, Director, Curriculum Frameworks and Instructional Resources Division

Thomas Adams, Administrator, Curriculum Frameworks Unit, Curriculum Frameworks and Instructional Resources Division

Christopher Dowell, Education Programs Consultant, Curriculum Frameworks Unit

Rona Gordon, Education Programs Consultant, Child, Youth, and Family Services Branch

Diane Hernandez, Education Programs Consultant, Standards and Assessment Division

Barbara Jeffus, School Library Consultant, Curriculum Frameworks Unit

Phil Lafontaine, Education Programs Consultant, Professional Development and Curriculum Support Division

Martha Rowland, Education Programs Consultant, Curriculum Frameworks Unit

Stacy Sinclair, Education Programs Consultant, Curriculum Frameworks Unit

Mary Sprague, Education Programs Consultant, Education Technology Office

William Tarr, Education Programs Consultant, Standards and Assessment Division

Deborah Tucker, Education Programs Consultant, Professional Development and Curriculum Support Division

Christine Bridges, Analyst, Curriculum Frameworks Unit

Belen Mercado, Analyst, Curriculum Frameworks Unit

Teri Ollis, Analyst, Curriculum Frameworks Unit

Lino Vicente, Analyst, Curriculum Frameworks Unit

Tracie Yee, Analyst, Curriculum Frameworks Unit

Tonya Odums, Office Technician, Curriculum Frameworks Unit

Beverly Wilson, Office Technician, Curriculum Frameworks Unit

State Board of Education Policy on the Teaching of Natural Sciences

The domain of the natural sciences is the natural world. Science is limited by its tools—observable facts and testable hypotheses.

Discussions of any scientific fact, hypothesis, or theory related to the origins of the universe, the earth, and life (the *how*) are appropriate to the science curriculum. Discussions of divine creation, ultimate purposes, or ultimate causes (the *why*) are appropriate to the history–social science and English–language arts curricula.

Nothing in science or in any other field of knowledge shall be taught dogmatically. Dogma is a system of beliefs that is not subject to scientific test and refutation. Compelling belief is inconsistent with the goal of education; the goal is to encourage understanding.

To be fully informed citizens, students do not have to accept everything that is taught in the natural science curriculum, but they do have to understand the major strands of scientific thought, including its methods, facts, hypotheses, theories, and laws.

A scientific fact is an understanding based on confirmable observations and is subject to test and rejection. A scientific hypothesis is an attempt to frame a question as a testable proposition. A scientific theory is a logical construct based on facts and hypotheses that organizes and explains a range of natural phenomena. Scientific theories are constantly subject to testing, modification, and refutation as new evidence and new ideas emerge. Because scientific theories have predictive capabilities, they essentially guide further investigations.

From time to time natural science teachers are asked to teach content that does not meet the criteria of scientific fact, hypothesis, and theory as these terms are used in natural science and as defined in this policy. As a matter of principle, science teachers are professionally bound to limit their teaching to science and should resist pressure to do otherwise. Administrators should support teachers in this regard.

Philosophical and religious beliefs are based, at least in part, on faith and are not subject to scientific test and refutation. Such beliefs should be discussed in the social science and language arts curricula. The Board's position has been stated in the *History–Social Science Framework* (adopted by the Board).[1] If a student should raise a question in a natural science class that the teacher determines is outside the domain of science, the teacher should treat the question with respect. The teacher should explain why the question is outside the domain of natural science and encourage the student to discuss the question further with his or her family and clergy.

Neither the California nor the United States Constitution requires that time be given in the curriculum to religious views in order to accommodate those who object to certain material presented or activities conducted in science classes. It may be unconstitutional to grant time for that reason.

Nothing in the California *Education Code* allows students (or their parents or

Note: This policy statement on the teaching of natural sciences, which was adopted by the State Board of Education in 1989, supersedes the State Board's 1972 Antidogmatism Policy.

guardians) to excuse their class attendance on the basis of disagreements with the curriculum, except as specified for (1) any class in which human reproductive organs and their functions and process are described, illustrated, or discussed; and (2) an education project involving the harmful or destructive use of animals. (See California *Education Code* Section 51550 and Chapter 2.3 of Part 19 commencing with Section 32255.) However, the United States Constitution guarantees the free exercise of religion, and local governing boards and school districts are encouraged to develop statements, such as this one on policy, that recognize and respect that freedom in the teaching of science. Ultimately, students should be made aware of the difference between *understanding*, which is the goal of education, and *subscribing to* ideas.

Notes

1. *History–Social Science Framework for California Public Schools* (Updated edition with content standards). Sacramento: California Department of Education, 2001.

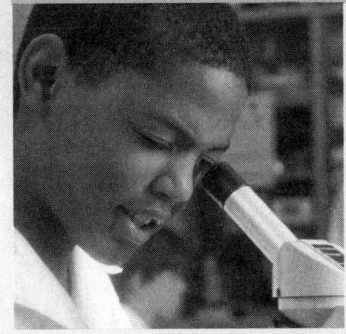

Introduction to the Framework

Introduction to the Framework

California is a world leader in science and technology and, as a result, enjoys both prosperity and a wealth of intellectual talent. The nation and the state of California have a history that is rich in innovation and invention. Educators have the opportunity to foster and inspire in students an interest in science; the goal is to have students gain the knowledge and skills necessary for California's workforce to be competitive in the global, information-based economy of the twenty-first century.

The *Science Framework for California Public Schools* is the blueprint for reform of the science curriculum, instruction, professional preparation and development, and instructional materials in California. The framework outlines the implementation of the *Science Content Standards for California Public Schools* (adopted by the State Board of Education in 1998)[1] and connects the learning of science with the fundamental skills of reading, writing, and mathematics. The science standards contain a concise description of what to teach at specific grade levels; this framework extends those guidelines by providing the scientific background and the classroom context.

Glenn T. Seaborg, one of the great scientific minds of this era, defined science as follows: "Science is an organized body of knowledge and a method of proceeding to an extension of this knowledge by hypothesis and experiment."[2] This framework is intended to (1) organize the body of knowledge that students need to learn during their elementary and secondary school years; and (2) illuminate the methods of science that will be used to extend that knowledge during the students' lifetimes.

Although the world will certainly change in ways that can hardly be predicted in the new century, California students will be prepared to meet new challenges if they have received a sound, basic education. This framework outlines the foundation of science knowledge needed by students and the analytical skills that will enable them to advance that knowledge and absorb new discoveries.

Audiences for the Framework

One of the primary audiences for this framework is the teachers who are responsible for implementing the science standards. These teachers are elementary and middle school teachers with multiple-subject credentials, middle and high school teachers with single-subject credentials in science, and those who may be teaching outside their primary area of expertise. The *Science Framework* is designed to provide valuable insights to both novice and expert science teachers.

For designers of science instructional materials, the framework may serve as a guide to the teaching of the science standards and as an example of the scholarly treatment of science that is expected in their materials. Publishers submitting science instructional materials for adoption in the state of California must adhere to a set of rigorous criteria described in this framework. The criteria include careful alignment with and comprehensive coverage of the science standards, good program organization and provisions for assessment, universal access for students with special needs, and instructional planning and support for the teacher.

The organizers of both programs of preservice professional preparation and in-service professional development will find this framework helpful. Skill is needed to teach science well, and training programs for teachers need to be especially mindful of the expectations placed on students.

Scientists and other professionals in the community often seek ways to help improve their local schools, and this framework will be helpful in focusing their efforts on a common set of curricular goals. By providing ideas and resources aligned with grade-level standards, professionals can make sure their outreach efforts and donations to classrooms will be put to best use.

For many high school seniors, commencement is followed shortly thereafter by baccalaureate courses. The *Science Framework* communicates to the science faculty at all California institutions of higher education what they may expect of entering students.

Finally, the parents, guardians, and other caregivers of students will find the *Science Framework* useful as they seek to help children with homework or gain an understanding of what their children are learning in school.

Instructional Materials

One of the best measures that local educational agencies (LEAs) can take is to ensure that all teachers of students in kindergarten through grade eight are provided with materials currently adopted by the State Board of Education (especially in science, mathematics, and reading/language arts) and are trained in their use. Those materials undergo a rigorous review and provide teachers and other instructional staff with guidance and strategies for helping students who are having difficulty.

In choosing instructional materials at the high school level, LEAs need to be guided by the science standards and the evaluation criteria set forth in Chapter 9. An analogy used in the *Reading/Language Arts Framework for California Public Schools* is equally applicable to the teaching of science: "Teachers should not be expected to be the composers of the music as well as the conductors of the orchestra."[3] In addition to basic instructional materials, teachers need to be able to gain access to up-to-date resources in the school library-media center that support the teaching of standards-based science. The resources must be carefully selected to support and enhance the basic instructional materials.

The Challenges in Science Education

Elementary school students often learn much from observing and recording the growth of plants from seeds in

Chapter 1
Introduction to the Framework

the classroom. But are the same students well served if seed planting is a focus of the science curriculum in the next year and the following one as well? The same question may well be asked of any instructional activity. To overcome the challenges in science education, several strategies are recommended:

Prepare Long-Term Plans

Long-term planning of a science curriculum over a span of grades helps students learn new things and develop new skills each year. A standards-based curriculum helps students who move from district to district; they will be more likely to receive a systematic and complete education.

The *Science Content Standards* and the *Science Framework* are designed to ensure that all students have a rich experience in science at every grade level and that curriculum decisions are not made haphazardly. Instructional programs need the content standards to be incorporated at each grade level and should be comprehensive and coherent over a span of grade levels.

Reforming science curriculum, instruction, and instructional materials will be a time-consuming process. To achieve the reform objectives, all educational stakeholders need to adhere to the guidance provided in this framework. The hope is that in the near future teachers will have a much greater degree of certainty about the knowledge and skills the students already possess as they file into the classroom at the beginning of a school year. Less time will be spent on review, and teachers will also have a clear idea of the content their students are expected to master at each grade level and in each branch of science.

Meet the Curricular Demands of Other Core Content Areas

The *Reading/Language Arts Framework* and *Mathematics Framework*[4] explicitly require uninterrupted instructional time in those subjects. In the early elementary grades, students need to receive at least 150 minutes of reading/language arts instruction daily and 50 to 60 minutes of mathematics instruction. At the elementary school level, the pressure to raise the academic performance of students in reading/language arts and mathematics has led some administrators to eliminate or curtail science instruction. This action is not necessary and reflects, in fact, a failure to serve the students. The *Science Framework* helps to organize and focus elementary science instruction, bringing it to a level of efficiency so that it need not be eliminated.

All teachers, particularly those who teach multiple subjects, need to use their instructional time judiciously. One of the key objectives set forth in the *Mathematics Framework* applies equally well to the study of science: "During the great majority of allocated time, students are active participants in the instruction."[5] In this case *active* means that students are engaged in thinking about science. If the pace of an activity is too fast or too slow, students will not be "on task" for much of the allotted time.

When large blocks of time for science instruction are not feasible, teachers must make use of smaller blocks. For example, an elementary teacher and the class may have a brief but spirited discussion on why plant seeds have different shapes or why the moon looks different each week. For kindergarten through grade three, standards-based science content is now integrated into nonfiction material in the basic reading/language arts reading programs adopted by the State Board of Education. Publishers were given the following mandate in the *2002 K–8 Reading/Language Arts/English Language Development Adoption Criteria:*

> In order to protect language arts instructional time, those K–3 content standards in history–social science and science that lend themselves to instruction during the language arts time period are addressed within the language arts materials, particularly in the selection of expository texts that are read to students, or that students read.[6]

There is no begrudging of the extended time needed for students to master reading, writing, and mathematics, for those are fundamental skills necessary for science. The *Reading/Language Arts Framework* states this principle clearly: "Literacy is the key to becoming an independent learner in all the other disciplines."[7] The *Mathematics Framework* bears a similar message: "The [mathematics] standards focus on essential content for all students and prepare students for the study of advanced mathematics, science and technical careers, and postsecondary study in all content areas."[8]

Despite the aforementioned curricular demands, the science standards should be taught comprehensively during the elementary grades. This challenge can be met with careful planning and implementation.

Set Clear Instructional Objectives

In teaching the science standards, LEAs must have a clear idea of their instructional objectives. Science education is meant to teach, in part, the specific knowledge and skills that will allow students to become literate adults. As John Stuart Mill wrote in 1867:

> It is surely no small part of education to put us in intelligent possession of the most important and most universally interesting facts of the universe, so that the world which surrounds us may not be a sealed book to us, uninteresting because unintelligible.[9]

Science education, however, is more than the learning of interesting facts; it is the building of intellectual strength in a more general sense:

> The scholarly and scientific disciplines won their primacy in traditional programs of education because they represent the most effective methods which . . . have been [devised] through millennia of sustained effort, for liberating and organizing the powers of the human mind.[10]

Science education in kindergarten through grade twelve trains the mind and builds intellectual strength and must not be limited to the lasting facts and skills that can be remembered into adulthood. Science must be taught at a level of rigor and depth that goes well beyond what a typical adult knows. It must be taught "for the sake of science" and not with any particular vocational goal in mind. The study of science disciplines the minds of students; and the benefits of this intellectual training are realized long after schooling, when the details of the science may be forgotten.

Model Scientific Attitudes

Science must be taught in a way that is scholarly yet engaging. That is, an appropriate balance must be maintained between the fun and serious sides of science. A physics teacher might have students build paper airplanes to illustrate the relationship between lift and drag in airflow; but if the activity is not deeply rooted in the content of physics, then the fun of launching paper airplanes displaces the intended lesson. The fun of science may be a way to help students remember important ideas, but it cannot substitute for effective instruction and sustained student effort.

There are certain attitudes about science and scientists that a teacher must foster in students. Scientists are deeply knowledgeable about their fields of study but typically are willing to admit that there is a great deal they do not know. In particular, they welcome new ideas that are supported by evidence. In doing their research good scientists do not attempt to prove that their own hypotheses are correct but that they are incorrect.[11] Though somewhat counterintuitive, this path is the surest one to finding the truth.

Classroom teachers must always provide rational explanations for phenomena, not occultic or magical ones. They need to be honest about what they do not know and be enthusiastic about learning new things along with their students. They must convey to students the idea that there is much to learn and that phenomena not currently understood may be understood in the future. Knowledge in science is cumulative, passed from generation to generation, and refined at every step.

Provide Balanced Instruction

Some of the knowledge of science is best learned by having students read about the subject or hear about it from the teacher; other knowledge is best learned in laboratory or field studies. Direct instruction and investigative activities need to be mutually supportive and synergistic. Instructional materials need to provide teachers with a variety of options for implementation that are based on the science standards.

For example, students might learn about Ohm's law, one of the guiding principles of physics, which states that electrical current decreases proportionately as resistance increases in an electrical circuit operating under a condition of constant voltage. In practice, the principle accounts for why a flashlight with corroded electrical contacts does not give a bright beam, even with fresh batteries. It is a simple relationship, expressed as *V=IR,* and embodied in high school Physics Standard 5.b. In a laboratory exercise, however, students may obtain results that seem to disprove the linear relationship because the resistance of a circuit element varies with temperature. The temperature of the components gradually increases as repeated tests are performed, and the data become skewed.

In the foregoing example, it was not Ohm's law that was wrong but an assumption about the stability of the experimental apparatus. This assumption can be proven by additional experimentation and provides an extraordinary opportunity for students to learn about the scientific method.

Had the students been left to uncover on their own the relationship between current and resistance, their skewed data would not have easily led them to discover Ohm's law. A sensible balance of direct instruction and investigation and a focus on demonstration of scientific principles provide the best science lesson.

Ensure the Safety of Instructional Activities

Safety is always the foremost consideration in the design of demonstrations, hands-on activities, laboratories, and science projects on site or away from school. Teachers need to be familiar with the *Science Safety Handbook for California Public Schools.*[12] It contains specific and useful information relevant to classroom teachers of science. Following safe practices is a legal and moral obligation for administrators, teachers, parents or guardians, and students. Safety needs to be taught. Scientists and engineers in universities and industries are required to follow strict environmental health and safety regulations. Knowing and following safe practices in science are a part of understanding the nature of science and scientific procedures.

Match Instructional Activities with Standards

Teachers need to use instructional materials that are aligned with the *Science Content Standards,* but how do they know when a curriculum or supplemental material is a good match? The State Board of Education establishes content review panels to analyze the science instructional materials submitted for adoption in kindergarten through grade eight. The panels consist of professional scientists and expert teachers of science. Local educational agencies would be well advised to use materials that have passed this stringent test for quality and alignment. The criteria are included in the *Science Framework* (Chapter 9) and may help guide the decisions of school districts and schools when they adopt instructional materials for grades nine through twelve.

In brief, teachers need to use instructional activities or readings that are grounded in science and that provide clear and nonsuperficial lessons. The content must be scientifically accurate, and the breadth and depth of the science standards need to be addressed. Initial teaching sequences must communicate with students in the most straightforward way possible, and expanded teaching used to amplify the students' understanding.[13] The concrete examples, investigative activities, and vocabulary used in instruction need to be unambiguous and chosen to demonstrate the wide range of variation on which scientific concepts can be generalized.

For example, in grade four Standard 2.a is: "Students know plants are the primary source of matter and energy entering most food chains."[14] This standard may be taught by using numerous concrete examples. Mastery of the concept, however, requires that students understand how the concept is generalized. Having learned by explicit instruction that plants are primary producers in deserts, forests, and grasslands, the students must be able to generalize the principle accurately to include other habitats (e.g., salt marshes, lakes, tundra). Although the standard is easily amenable to laboratory and field activities, it cannot be entirely grasped by observation of or contact with nature.

In high school this standard is explored in considerable depth as students come to learn about energy, matter, photosynthesis, and the cycling of organic matter in an ecosystem. Standard 2.a in grade four prepares students to learn more.

Another example of "preteaching" embedded in the science standards is Standard 1.h in grade three: "Students know all matter is made of small particles called atoms, too small to see with the naked eye."[15] The intent of this standard is not to make third-grade students into atomic scientists, but simply to introduce them to a way of thinking that is reinforced in grades five and eight and then taught in greater depth in high school.

This framework is designed to ensure that instructional materials are developed to the intended depth of each standard and that the relationships are made clear among standards across grade levels and within branches of science.

Science and the Environment

Environmental concerns that once received relatively little attention (e.g., invasive species of plants and animals, habitat fragmentation, loss of biodiversity) have suddenly become statewide priorities. Entire fields of scientific inquiry (e.g., conservation biology, landscape ecology, ethnoecology) have arisen to address those concerns. In general, there is an increased sense of the complexity and interconnectedness of environmental issues. The public response to California's environmental challenges has been profound as evidenced by the enactment of Senate Bill 373 (Chapter 926, Statutes of 2001). Senate Bill 373 requires the following topics to be included in this framework:

- Integrated waste management
- Energy conservation
- Water conservation and pollution prevention
- Air resources
- Integrated pest management
- Toxic materials
- Wildlife conservation and forestry

Several science standards address those topics directly; provide students with the foundational skills and knowledge to understand them; or incorporate concepts, principles, and theories of science that are integral to them. The suggestions in this framework include ways of highlighting the topics as follows:

- Students in kindergarten through grade five learn about the characteristics of their environment through their studies of earth, life, and physical sciences. For example, in grade three students learn how environmental changes affect living organisms.

- Students in grades six through eight focus on earth, life, and physical sciences, respectively; and standards at each grade level include the study of ecology and the environment.

- Students in grades nine through twelve expand their knowledge of habitats, biodiversity, and ecosystems associated with the biology/life science content standards. High school earth science standards include the study of energy and its usage as well as topics related to water resources and the geology of California.

The Legislature has declared "that [we have] a moral obligation to understand the world in which [we live] and to protect, enhance, and make the highest use of the land and resources [we hold] in trust for future generations, and that the dignity and worth of the individual requires a quality environment in which [we] can develop the full potentials of [our] spirit and intellect" (*Education Code* Section 8704). Toward that end LEAs and individual schools throughout California are contributing to the betterment of the environment in many ways, including replacing asphalt school grounds with gardens, recycling school waste, exchanging scientific data with the international community through Web sites, and restoring local habitats.

Specific programs of environmental education enhance the learning of science at all grade levels. These programs enhance scientific and critical thinking skills, enabling students to perceive patterns and processes of nature, research environmental issues, and propose reasoned solutions. Environmental education is not advocacy for particular opinions or interests, but it is a means of fostering a comprehensive and critical approach to issues. Students get a personal sense of responsibility for the environment; consequently, schools are tied more closely to the life of the communities they serve.

Guiding Principles

The following principles form the basis of an effective science education program. They address the complexity of the science content and the methods by which science content is best taught. They clearly define the attributes of a quality science curriculum at the elementary, middle, and high school levels.

Effective science programs are based on standards and use standards-based instructional materials.

Comprehensive, standards-based programs are those in which curriculum, instruction, and assessment are aligned with the grade level-specific content standards (kindergarten through grade eight) and the content strands (grades nine through twelve). Students have opportunities to learn foundational skills and knowledge in the elementary and middle grades and to understand concepts, principles, and theories at the high school level. Students use instructional materials that have been adopted by the State Board of Education in kindergarten through grade eight. For grades nine through twelve, students use instructional materials that are determined by local boards of education to be consistent with the science standards and this framework.

A *California Standards Test* in science is now administered at grade five, reflecting the cumulative science standards for grades four and five. Therefore, science instruction must be based on complete programs that cover all the standards at every grade level. The criteria for evaluating K–8 science instructional materials (see Chapter 9) state: "All content Standards as specified at each grade level are supported by topics or concepts, lessons, activities, investigations, examples, and/or illustrations, etc., as appropriate."

At the high school level, the *Science Content Standards* document does not prescribe a single high school curriculum. To allow LEAs and teachers flexibility, the standards for grades nine through twelve are organized as content strands. There is no mandate that a particular content strand be completed in a particular grade. Students enrolled in science courses are expected to master the standards that apply to the curriculum they are studying regardless of the sequence of the content. Students in grades nine through twelve use instructional materials that reflect the *Science Content Standards* and this framework. The grade nine through twelve standards maps posted on the California Department of Education Web site are tools that LEAs can use to determine if instructional materials are aligned with the standards.

The *California Standards Tests* for grades nine through eleven pertain specifically to the content of the particular science courses in which students are enrolled. The California Department of Education makes blueprints for those tests and sample questions available to the public. Local educational agencies are encouraged to review and improve (as necessary) their high school science programs to achieve the following results:

1. All high school science courses that meet state or local graduation requirements or the entrance requirements of the University of California or The California State University are based on the *Science Content Standards*.

2. Every laboratory science course is based on the content standards and ensures that students master both the content-specific standards and Investigation and Experimentation standards.

3. Every science program ensures that students are prepared to be successful on the *California Standards Tests*.

4. All students take, at a minimum, two years of laboratory science providing fundamental knowledge in at least two of the following content strands: biology/life sciences, chemistry, and physics. Laboratory courses in earth sciences are acceptable if prerequisite courses are required (or provide basic knowledge) in biology, chemistry, or physics.[16]

Effective science programs develop students' command of the academic language of science used in the content standards.

The lessons explicitly teach scientific terms as they are presented in the content standards. New words (e.g., *photosynthesis*) are introduced to reflect students' expanding knowledge, and the definitions of common words (e.g., *table*) are expanded to incorporate specific meanings in science. Developing students' command of the academic language of science must be a part of instruction at all grade levels (kindergarten through grade eight) and in the four content strands (grades nine through twelve). Scientific vocabulary is important in building conceptual understanding. Teachers need to provide explanations of new terms and idioms by using words and examples that are clear and precise.

Effective science programs reflect a balanced, comprehensive approach that includes the teaching of investigation and experimentation skills along with direct instruction and reading.

A balanced, comprehensive approach to science includes the teaching of investigation and experimentation skills along with direct instruction and reading. Investigation and experimentation standards are progressive and need to be taught in a manner integral to the physical, life, and earth sciences as students learn quantitative skills and qualitative observational skills. For example, the metric system is first introduced in grade two, but students use and refine their skill in metric measurement through high school. The methods and skills of scientific inquiry are learned in the context of the key concepts, principles, and theories set forth in the standards. Effective use of limited instructional time is always a major consideration in the design of lessons and courses. Laboratory space and equipment, library access, and resources are essential to support students' academic growth in science.

Effective science programs use multiple instructional strategies and provide students with multiple opportunities to master the content standards.

Multiple instructional strategies, such as direct instruction, teacher modeling and demonstration, and investigation and experimentation, are useful in teaching science and need to be included in instructional materials. Those strategies help teachers capture student interest, provide bridges across content areas, and contribute to an understanding of the nature of science and the methods of scientific inquiry.

Standards for investigation and experimentation are included at each grade level and differ from the other standards in that they do not represent a specific content area. Investigation and experimentation cuts across all content areas, and those standards are intended to be taught in the context of the grade-level content. Hands-on activities compose at least 20 to 25 percent of the science instructional time in kindergarten through grade eight. Instruction is designed and sequenced to provide students with opportunities to reinforce foundational skills and knowledge and to revisit concepts, principles, and theories previously taught. In this way student progress is appropriately monitored.

Effective science programs include continual assessment of students' knowledge and understanding, with appropriate adjustments being made during the academic year.

Effective assessment (on a continuing basis through the academic year) is a key ingredient of standards-based instruction. Teachers assess students' prerequisite knowledge, monitor student progress, and evaluate the degree of mastery of the content called for in the standards. Lessons include embedded unit assessments that provide formative and summative assessments of student progress. Teachers and administrators regularly collaborate to improve science progress by examining the results of *California Standards Tests* in science (both the general test at grade five and the specific tests in grades nine through eleven).

 **Effective science programs continually engage all students in learning and prepare and motivate students for further instruction in science.**

Students who are unable to keep up with the expectations for learning science often lack basic skills in reading comprehension and mathematics. Therefore, students who need extra assistance to achieve grade-level expectations are identified early and receive support. Schools need to use transitional materials that accelerate the students' reading and mathematics achievement to grade level. Advanced learners must not be held back but be encouraged to study science content in greater depth.

 **Effective science programs use technology to teach students, assess their knowledge, develop information resources, and enhance computer literacy.**

Across the nation science in the laboratory setting involves specialized probes, instruments, materials, and computers. Scientists extend their ability to make observations, analyze data, study the scientific literature, and communicate findings through the use of technology. High-performance computing capabilities are used in science to make predictions based on fundamental principles and laws. Technology-based models are used to design and guide experiments, making it possible to eliminate some experiments and to suggest other experiments that previously might not have been considered. Students have the opportunity to use technology and imitate the ways of modern science. Teaching science by using technology is important for preparing students to be scientifically and technologically literate. Assembly Bill 1023 (Chapter 404, Statutes of 1997) requires that newly credentialed teachers demonstrate basic competence in the use of computers in the classroom.

 **Effective science programs have adequate instructional resources as well as library-media and administrative support.**

Standards-based teaching and learning in science demand adequate instructional resources. Local educational agencies and individual school sites need to include science resources as an integral part of the budget. Library-media staff must have science as a priority for resource acquisition and development. Administrators must ensure that funds set aside for the science resources are spent efficiently (e.g., through clear processes and procedures for purchasing and maintenance) and support students' mastery of the content standards. This priority requires planning, coordination, and dedication of space for science resources.

 **Effective science programs use standards-based connections with other core subjects to reinforce science teaching and learning.**

Science instruction provides multiple opportunities to make connections with other content areas. Reading, writing, mathematics, and speaking skills are needed to learn and do science. In self-contained classrooms, teachers incorporate science content in reading, writing, and mathematics as directed in the *Reading–Language Arts Framework* and *Mathematics Framework*. In departmentalized settings (middle and high school levels) science teachers need to include essay assignments and require that students' writing reflect the correct application of English-language conventions, including spelling and grammar.

Organization of the Framework

The *Science Framework* is primarily organized around the *Science Content Standards*. The framework:

- Discusses the nature of science and technology and the methods by which they are advanced (Chapter 2)
- Describes the curriculum content and instructional practices needed for mastery of the standards (Chapters 3, 4, and 5)
- Guides the development of appropriate assessment tools (Chapter 6)
- Suggests specific strategies to promote access to the curriculum for students with special needs (Chapter 7)
- Describes the system of teacher professional development that needs to be in place for effective implementation of the standards (Chapter 8)
- Specifies the requirements for evaluating science instructional resources, including investigative activities, for kindergarten through grade eight (Chapter 9)
- Provides information on pertinent requirements of the California *Education Code* regarding science education in this state (Appendix)

The science standards are embedded in Chapters 3, 4, and 5 and are grade-level specific from kindergarten through grade eight. The standards for grades nine through twelve are organized by strands: physics, chemistry, biology/life sciences, and earth sciences.

Notes

1. *Science Content Standards for California Public Schools, Kindergarten Through Grade Twelve.* Sacramento: California Department of Education, 2000.
2. Glenn T. Seaborg, "A Letter to a Young Scientist," in *Gifted Young in Science: Potential Through Performance.* Edited by Paul Brandwein and others. Arlington, Va.: National Science Teachers Association, 1989. The late Dr. Seaborg was chair of the California Academic Standards Commission's Science Committee that created the *Science Content Standards for California Public Schools.*
3. *Reading/Language Arts Framework for California Public Schools, Kindergarten Through Grade Twelve.* Sacramento: California Department of Education, 1999, p. 2.
4. *Mathematics Framework for California Public Schools, Kindergarten Through Grade Twelve.* Sacramento: California Department of Education, 2000.
5. *Mathematics Framework*, p. 13.
6. These criteria are available on the Web site <http://www.cde.ca.gov/cfir/rla/2002criteria.pdf>. Click on Criteria Category 2, third bullet.
7. *Reading/Language Arts Framework*, p. 3.

8. *Mathematics Framework,* p. 18.
9. John Stuart Mill, inaugural address to the University of St. Andrew, quoted in George E. DeBoer, *A History of Ideas in Science Education.* New York: Teachers College Press, 1991, p. 8.
10. Arthur E. Bestor, *Educational Wastelands: The Retreat from Learning in Our Public Schools.* Champaign: University of Illinois Press, 1953. p. 18.
11. J. R. Platt, "Strong Inference," *Science,* Vol. 146 (1964), 347–53.
12. *Science Safety Handbook for California Public Schools.* Sacramento: California Department of Education, 1999.
13. Siegfried Engelmann and Douglas Carnine, *Theory of Instruction: Principles and Applications.* Eugene, Ore.: ADI Press, 1991.
14. *Science Content Standards,* p. 11.
15. Ibid., p. 8.
16. The laboratory science subject requirement for admission to the University of California and (beginning in fall 2003) to The California State University reads as follows: "d. Laboratory Science. Two years required, three recommended. Two years of laboratory science providing fundamental knowledge in at least two of these three disciplines: biology (which includes anatomy, physiology, marine biology, aquatic biology, etc.), chemistry, and physics. Laboratory courses in earth/space sciences are acceptable if they have as prerequisites or provide basic knowledge in biology, chemistry, or physics. The appropriate two years of an approved integrated science program may be used to fulfill this requirement. Not more than one year of ninth-grade laboratory science can be used to meet this requirement."

Source: University of California Office of the President <*http://www.ucop.edu*> and The California State University <*http://www.calstate.edu*>.

2

The Nature of Science and Technology

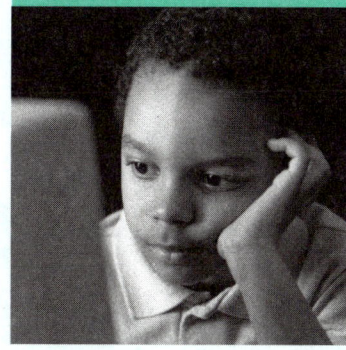

The Nature of Science and Technology

Science is the study of nature at all levels, from the infinite to the infinitesimal. It is the asking and answering of questions about natural processes or phenomena that are directly observed or indirectly inferred. From these questions and answers come tentative explanations called *hypotheses,* which lead to testable predictions about the natural processes and phenomena. Hypotheses that withstand rigorous testing of predictions will gradually lead to an accretion of facts and principles, which serve as the foundation of scientific theories. Scientific knowledge gives rise to many technologies that drive the economy and improve the quality of life for people. The term *technology* embraces not only tools (e.g., computers) but also methods, materials, and applications of scientific knowledge.

Scientific knowledge and technology built on that knowledge have expanded—one might even say exploded—in the last 50 years. The very nature of science as a human endeavor has made this expansion possible. Scientific research and development are both collaborative and international; literally millions of men and women around the world participate in the science and engineering enterprise.

To stay current with scientific developments, school science programs need to develop partnerships with library-media centers, museums, science and technology centers, colleges and universities, industry, and subject matter projects to build support for such programs.

The Scientific Method

The scientific method is a process for predicting, on the basis of a handful of scientific principles, what will happen next in a natural sequence of events. Because of its success, this invention of the human mind is used in many fields of study. The scientific method is a flexible, highly creative process built on three broad assumptions:

- Change occurs in observable patterns that can be extended by logic to predict what will happen next.
- Anyone can observe something and apply logic.
- Scientific discoveries are replicable.

The first assumption may be contrasted with the idea that the complexity of the natural world is so great as to be outside human understanding. Science asserts that change occurs in patterns within the human capability to perceive, that these patterns may be discerned by observation, and that the changes are subject to logic. The simple logic of *if-A-then-B* suffices to understand simple patterns; complex patterns require the more complex logic expressed in mathematics. The concentration of thought that mathematics lends to the study of the natural world is stunning. For example, the position of the planets over millions of years in the past can be closely approximated by using only two strings of symbols taken from Newton's laws of motion and gravity:

$F = ma$ and $F = Gm_1 m_2 / d^2$

The second assumption is that anyone can measure the strength of a scientific theory by fairly testing its predictions. Scientific research papers thus contain not only the results of an investigation, but all the information needed to replicate the research. The lifetime work of many scientists is replicating other scientists' experiments in order to test their conclusions.

The third assumption is that individuals and groups can make progress toward understanding natural phenomena, and their discoveries can be replicated at any time and suitable place by an objective observer. Truth in science knows no cultural or national boundaries. Science is not a system of beliefs or faith but a replicable body of knowledge. In fact, science is incapable of answering questions that are based on faith. Within the scientific community, individuals or groups may sometimes see only what they desire to see or have been conditioned to see. Scientific progress is sometimes stalled by incorrect theories or results; but once it is shown that those theories or results cannot be confirmed by others, progress resumes in the correct direction.

The scientific method ultimately allows for the formulation of scientific theories. Part of science education is to learn what these theories are and trace their operation in the world. A *theory* in popular language is a collection of related ideas that one supposes to be true; in science, a theory is defined by the principles of the scientific method. Those principles, in order of precedence, are as follows:

1. A scientific theory must be logically consistent and lead to testable predictions about the natural world.
2. The strength of a scientific theory lies solely in the accuracy of those predictions.
3. Of two scientific theories that make accurate predictions, the theory that makes a greater number of predictions with fewer underlying assumptions is likely to prove stronger.

The making and testing of predictions is what distinguishes science from other intellectual disciplines, and emphasizing the accuracy of predictions rather than the cogency of explanations is the key to scientific progress. A large assortment of recorded observations can often be accounted for with explanations that sound good but are nonetheless wrong. Predictions that can be tested and verified, however, provide a sound standard by which a scientific theory can be judged.

The requirement that a scientific theory makes predictions might seem to reject as unscientific any theory that describes the past. However, scientific theories that describe the past (such as those set forth by geologists, paleontologists, and so forth) do make predictions about what will be observed in the future. For example, if there was a mass extinction at the boundary between the Cretaceous and Tertiary periods more than 65 million years ago, then a sample of sedimentary rock at or below that boundary would be expected (at many sites) to show a much greater number (and diversity) of fossils than would a sample of rock immediately above it. Furthermore, if the extinction were to have been caused by the impact of an asteroid, the layer of "dust" falling after the impact would be expected to show signs of the explosive nature of the impact and the composition of the asteroid.

The three principles of the scientific method cannot be used to determine how the predictions of scientific theory should be tested. Nor can they be used to create new and even stronger theories. Inventing ways to test existing theories or devising new theories requires creativity as well as knowledge. The opportunity to make discoveries through creative experiments is what attracts many young people to science careers. As with all endeavors, practice, experience, and the opportunity to watch others engaged in similar efforts are highly beneficial to students.

Scientific Practice and Ethics

Scientists have the responsibility to report fully and openly the results of their experiments even if those results disagree with their favored hypothesis. They also have the responsibility to report fully and openly the methods of an experiment. For a scientist to hide data, arbitrarily eliminate anomalies in a data set, or conceal how an experiment was conducted is to invite errors and make those errors difficult to discover. All scientists must seek explanations for anomalous observations and results in order to improve their procedures or to discover something new. They also carefully consider questions raised by fellow scientists about the accuracy of their experiments. These accepted ethical practices of scientists need to be taught in the classroom at all grade levels.

Students sometimes feel pressure to come up with the right or expected answer when performing an investigation. Some may even alter the results of an experiment because they assume an error has been made in observing, measuring, or recording data. Teachers must encourage students to report the results they actually get. Variations in results allow students the opportunities to find problems with a procedure or an apparatus. Learning that such problems come with doing science and learning how to detect and correct these problems are as important as reaching the nominal goal of an investigation. To discard the unusual in order to reach the expected is to guarantee that nothing but the expected will ever be seen. That step would be a distortion of science.

Science and Technology

Technology is the repeatable and controlled manipulation of the natural world to serve human ends. A deep understanding of a phenomenon is not necessary for successful technology. For example, the Romans did not understand why mixtures of limestone powder, clay, sand, and water hardened, but they had no trouble using the result to mortar bricks. Still, the most spectacular technological advances often have followed from an understanding of fundamental scientific principles. Ancient peoples, for example, used the principles of genetic heredity to improve a few characteristics of some crops and animals, but not until Gregor Mendel's experiments on garden peas in the mid-1800s were the patterns of heredity discovered. Mendel's experiments marked the birth of genetic science, which has led to today's high-yield agricultural technologies and a new era in medicine. The Roman civilization and other earlier civilizations had the ability to combine and utilize raw materials. Only after scientists had sufficient knowledge of physics and chemistry to take those raw materials apart and rebuild them on a molecule-by-molecule (or even an atom-by-atom) basis did the stuff of today's technologies—from semiconductors and superconductors to a bewilderingly diverse array of synthetics—come into being.

As new scientific principles are discovered, new technologies can be devised, and the use of existing technologies can be expanded. The new technologies available even in the next decade cannot be predicted, but new technologies will certainly be needed to cope with foreseeable problems, such as human population growth, environmental pollution, and finite energy resources. It is also certain that technological advancements will be needed for the problems that no one has foreseen.

Teachers have the opportunity to incorporate technology, help students master the science standards, enhance students' abilities to use technology effectively, and help students understand the relationship between science and technology. Technology may serve the following functions in education:

- Enable teachers and students to have access to the latest information in science. In grade seven of the science standards, Standard 7.b calls for students to "use a variety of print and electronic resources (including the World Wide Web) to collect information and evidence as part of a research project."[1] Any Internet usage, of course, must comply with the provisions of applicable law and policies adopted by the local educational agency (LEA).

- Provide students with experience in effective communication. The investigation and experimentation standards provide a range of skills beginning in kindergarten ("Communicate observations orally and through drawings")[2] and culminating in grades nine through twelve ("Select and use appropriate tools and technology [such as computer-linked probes, spreadsheets, and graphing calculators] to perform tests, collect data, analyze relationships, and display data").[3]

- Further scientific study in business, industry, and postsecondary education. The technologies used to address energy needs are discussed in Chapter 4 under "Grade Six," and chips and semiconductors are discussed under "Grade Eight." In the high school grades, the science standards call for students to "know how genetic engineering (biotechnology) is used to produce novel biomedical and agricultural products."[4]

- Become integrated in instruction where it is likely to improve student learning. The focus must be on learning science and using technology as a tool rather than as an end in itself.

The *Mathematics Framework* contains an important precaution that equally applies to science: "The use of technology in and of itself does not ensure improvements in student achievement, nor is its use necessarily better for student achievement than are more traditional methods."[5]

- Simulate or model investigations and experiments that would be too expensive, time-consuming, dangerous, or otherwise impractical. Investigation and Experimentation Standard 1.g requires students in grades nine through twelve to "recognize the usefulness and limitations of models and theories as scientific representations of reality."[6]

- Support universal access to science content through assistive technologies, consistent with a student's 504 accommodation plan[7] or individualized education program.[8]

Resources for Teaching Science and Technology

As students learn the skills and knowledge called for in the *Science Content Standards for California Public Schools* they will come to know implicitly the nature of science. They will understand the key concepts, principles, and theories of science and will have practiced scientific inquiry. The guidance and information provided in this framework may be used to implement effective science education programs in public schools from kindergarten through grade twelve that will provide students with the opportunity to become scientifically literate and understand the nature of science and technology.

The number of electronic resources for science education is increasing rapidly. The California Learning Resource Network <http://www.clrn.org/science> provides a way for educators to identify supplemental electronic learning resources, including Web sites, that simultaneously meet local instructional needs and align with the *Science Content Standards* and this framework. Educators should comply with applicable policies of their school districts regarding Internet resources.

When science and technology are discussed in the context of history and historical figures, instruction is often enriched. The *History–Social Science Content Standards* follows this principle.[9] The standards:

- Include references to scientists such as Ben Franklin, Louis Pasteur, George Washington Carver, Marie Curie, Albert Einstein, Nicolaus Copernicus, Galileo Galilei, Johannes Kepler, and Isaac Newton.

- Cover inventors such as Thomas Edison, Alexander Graham Bell, the Wright brothers, James Watt, Eli Whitney, and Henry Bessemer.

- Encompass the effects of the information and computer revolutions, changes in communication, advances in medicine, and improvements in agricultural technology.

Science and Society

Science does not take place in a secret place isolated from the rest of society. Nor are the technologies that it creates shielded from public scrutiny. The continued expansion of scientific knowledge and the new technologies that spin off that knowledge will inevitably challenge citizens to rethink their ideas and beliefs. For example, as new genetically modified crops and livestock are developed by scientists, some people have expressed concern about food safety and the ethics of such practices. On the other hand, people in developing countries have a compelling interest to use the new technologies to rid themselves of famine and diseases. Those types of trade-offs are likely to become the focus of intense public discussion and political debate.

The presentation of some scientific findings or practices may be troubling to students who genuinely believe that those findings or practices conflict with their religious or philosophical beliefs. Dealing constructively and respectfully with those beliefs while holding firm to the nature of science is one of the greatest challenges to public school teachers.

Scientifically literate students need to understand clearly the major scientific theories and the principles behind the scientific method. They must also understand that though the scientific method is a powerful process for predicting natural phenomena, it cannot be used to answer moral and aesthetic questions. Nor can it be used to test hypotheses based on supernatural intervention. Science exclusively concerns itself with predicting the occurrence and consequences of natural events. This concern is explicitly expressed in Standard 7.10 of grade seven of the *History–Social Science Content Standards:* "Students analyze the historical developments of the Scientific Revolution and its lasting effect on religious, political, and cultural institutions"[10] The students go on to consider this analysis in terms of the roots of the scientific revolution, the significance of new scientific theories, the influence of new scientific rationalism on the growth of democratic ideas, and the coexistence of science with traditional religious beliefs.

Notes

1. *Science Content Standards for California Public Schools, Kindergarten Through Grade Twelve.* Sacramento: California Department of Education, 2000, p. 25.
2. Ibid., p. 2.
3. Ibid., p. 52.
4. Ibid., p. 44.
5. *Mathematics Framework for California Public Schools, Kindergarten Through Grade Twelve.* Sacramento: California Department of Education, 2000, p. 227.
6. *Science Content Standards,* p. 52.
7. A Section 504 accommodation plan is a document typically produced by school districts in compliance with the requirements of Section 504 of the federal Rehabilitation Act of 1973. The plan specifies agreed-on services and accommodations for a student who, as the result of an evaluation, is determined to have a "physical or mental impairment [that] substantially limits one or more major life activities." In contrast to the Individuals with Disabilities Education Act (IDEA), Section 504 allows a wide range of information to be contained in a plan: (1) the nature of the disability; (2) the basis for determining the disability; (3) the educational impact of the disability; (4) necessary accommodations; and (5) the least restrictive environment in which the student may be placed.

8. An individualized education program (IEP) is a written, comprehensive statement of the educational needs of a child with a disability and the specially designed instruction and related services to be employed to meet those needs. An IEP is developed (and periodically reviewed and revised) by a team of individuals, including the parent(s) or guardian(s), knowledgeable about the child's disability. The IEP complies with the requirements of the federal IDEA and covers such items as (1) the child's present level of performance in relation to the curriculum; (2) measurable annual goals related to involvement and progress in the curriculum; (3) specialized programs (or program modifications) and services to be provided; (4) participation with nondisabled children in regular classes and activities; and (5) accommodation and modification in assessments.

9. *History–Social Science Content Standards for California Public Schools, Kindergarten Through Grade Twelve.* Sacramento: California Department of Education, 2001.

10. Ibid., p. 31.

The Science Content Standards for Kindergarten Through Grade Five

3

The Science Content Standards for Kindergarten Through Grade Five

This chapter incorporates the *Science Content Standards,* providing an explanation of the science underlying the standards and outlining activities that are consistent with the objectives of the standards. The activities included in this chapter are examples of the ways in which the standards may be approached. They are not to be interpreted as requirements for the science classroom or for inclusion in instructional materials, thus explaining the frequent use of helping verbs, such as *can, may,* and *should.*

The science standards are set forth in terms of what students know. Therefore, mastery of an individual standard is achieved when students have actually learned the fact, skill, concept, principle, or theory specified. Mastery does not occur simply because students have received a particular explanation or participated in a particular activity.

The elementary school science program provides the foundational skills and knowledge students will need in middle school and high school. Students are introduced to facts, concepts, principles, and theories organized under the headings of physical, life, and earth sciences. They learn essential investigation and experimentation skills that will continue to be developed through high school. Elementary school students respond positively to well-structured activities and expository reading materials that connect the world around them to the science content. Students raise questions, follow their curiosity, and learn to be analytical. They are encouraged to practice open and honest expression of ideas and observations; they learn to listen to and consider the ideas and observations of other students. Both teachers and students need to enjoy the adventure of science.

This enjoyable adventure includes the school library-media center as a natural partner in science teaching and learning. The books and other resources available in the school library enhance and expand an interest in and understanding of science. When the school library-media center is appropriately staffed with a credentialed library-media teacher, information literacy instruction can be integrated into regular science instruction.

Safety is always the foremost consideration in teacher modeling and the design of demonstrations, investigation and experimentation, and science projects, both at the school site and away from school. Teachers must become familiar with the *Science Safety Handbook for California Public Schools.*[1] It contains specific and useful information relevant to classroom teachers of science. School administrators, teachers, parents/guardians, and students have a legal moral obligation to promote safety in science education. Safety must be taught. Scientists and engineers in universities and industries are required to follow strict environmental health and safety regulations. Knowing and following safe practices in science are a part of understanding the nature of science and scientific enterprise.

Kindergarten Science Content Standards

Science study provides children in kindergarten with a unique opportunity to explore the world around them. It is important to teach kindergarten students to be objective observers and to know the difference between an observation and an opinion. Students begin their study of science by observing and noting the similarities, differences, and component parts of materials, plants and animals, and the earth. They also observe processes and changes over time. Observational activities must always be designed with safety as a foremost consideration.

Students learn how to classify, compare, sort, and identify common objects. They expand their skills in descriptive language by learning to observe, measure, and predict the properties of materials. Activities related to freezing, melting, and evaporation can provide ways to stimulate classroom discussions. Studies of plants and animals, landforms, and weather allow students to recount personal stories and speak of familiar experiences and interests. In doing so they learn new vocabulary and have opportunities to practice mathematics. In the kindergarten curriculum, as students listen to stories, teachers may use important strategies for teaching comprehension by (1) using pictures and context to make predictions; (2) retelling familiar stories; and (3) answering and asking questions about essential elements.

STANDARD SET 1. Physical Science

Standard Set 1 begins the study of the properties of matter and its transformations. While learning these standards, students build a foundation for making observations and measurements. The three standards call attention to the properties of common objects (most of which are solids) and to the properties of water. Teachers introduce the term *physical property* to students by asking them to observe the properties of a variety of objects. Students will be able to predict on the basis of some initial observations what will happen under different conditions rather than make random guesses.

1. **Properties of materials can be observed, measured, and predicted. As a basis for understanding this concept:**
    a. *Students know* objects can be described in terms of the materials they are made of (e.g., clay, cloth, paper) and their physical properties (e.g., color, size, shape, weight, texture, flexibility, attraction to magnets, floating, sinking).

Students learn how to compare objects on the basis of characteristics and physical properties, such as color, size, shape, weight, texture, flexibility, attraction to magnets, and floating and sinking in water. By working with objects and noting their physical properties and characteristics, students develop their ability to make observations and use appropriate academic science language that is expressive and descriptive.

Teachers may provide a variety of objects that students can investigate by using the senses of sight, sound, and touch. Activities involving the sense of smell and taste should be done only at home under parental supervision. In the classroom students use sight, sound, and touch to sort objects according to their physical properties.

The next step is for students to sort objects according to properties that do not manifest themselves directly to those three senses. For example, they might test different objects for the ability to float or sink in a small container of water. The list might include wood blocks, sponges, solid rubber balls, metal washers, small rocks, and Styrofoam balls. Students can test a few of these objects by observing which ones sink or float, then test their predictions experimentally. They may be surprised to see that a heavy piece of wood will float, but a lighter metal washer will sink.

Those observations are important to discuss because the behavior of the object depends on its density and not its weight. Density is a topic that is covered formally in grade seven, but students need to get a "feel" for it in earlier grades. Similarly, the property of magnetism is discussed in some detail in grade four, but students in kindergarten may enjoy learning that magnets stick only to certain types of metal and that the most common magnetic metal is iron.

> **1. b.** *Students know* water can be a liquid or a solid and can be made to change back and forth from one form to the other.

Observing the change from ice to liquid water and back to ice builds students' understanding that a substance may have both solid and liquid forms. Freezing and then melting water shows students that the water is returned exactly to its original state. A teacher may consider reading a story about winter ice and snow to the class to help develop vocabulary and comprehension.

> **1. c.** *Students know* water left in an open container evaporates (goes into the air) but water in a closed container does not.

Students can observe a cup of water covered or uncovered in the classroom during several days or weeks. The gradual evaporation of water offers an opportunity for students to record observations and develop vocabulary related to time periods that extend beyond a single day. The rate of evaporation will depend on the temperature and room humidity and on the type of container chosen. Stories in which it rains and then the water dries up may also provoke interesting discussions. When rain forms puddles on the ground, some of the water may evaporate and some of it may sink into the soil where it can be taken up by plants.

Students may observe evaporation in classroom demonstrations. Water vapor in the air may be condensed to liquid water and collected on a cold surface. For example, a teacher might hold a hand-mirror over a container of hot tap water to show that the water vapor rises and fogs the mirror and that small droplets of water may form.

STANDARD SET 2. Life Sciences

Kindergarten students expand their observational skills and vocabulary by learning to describe the appearance and behavior of different animals and plants. They have the opportunity to discuss the principles of structure and function at a simple level. For example, most birds and many insects have wings and can fly, but birds have feathers and insects do not. There are many outstanding fictional stories that the teacher may select to read to the students, perhaps from the school library-media center. Students learn that stories may give plants and animals attributes that are funny but not real. Although authors may use anthropomorphism to engage the interest of young people and exercise their imagination, such a literary technique should not be confused with scientific procedures and fact. Teachers may use many expository texts to enrich the observation of plants and animals in a classroom.

> **2. Different types of plants and animals inhabit the earth. As a basis for understanding this concept:**
>
> a. *Students know* how to observe and describe similarities and differences in the appearance and behavior of plants and animals (e.g., seed-bearing plants, birds, fish, insects).

Teachers guide students to learn that all plants and animals need air, food, and water to grow and be healthy. Students also learn that most animals are able to move about from place to place, which helps them find food to eat. Terrestrial plants, on the other hand, are usually rooted in one place and must obtain their nutrients and energy from the surrounding air, soil, water, and sunlight.

> **2. b.** *Students know* stories sometimes give plants and animals attributes they do not really have.

Real plants and animals do not talk, wear clothing, or walk like humans. Scientific observation of plants and animals helps students in kindergarten to understand the difference between characteristics of the real world and of fantasy.

> **2. c.** *Students know* how to identify major structures of common plants and animals (e.g., stems, leaves, roots, arms, wings, legs).

Students increase the detail of their understanding of plants and animals as they learn about the major structural components of common plants and animals and their functions. For example, students might plant some seeds in pots, care for the plants that sprout, and note how the different structures (such as stems, leaves, and roots) change during growth and development. A comparison of different leaves is also instructive. Leaves that are good to study have smooth or jagged edges; are wide or narrow; and are of a different color, odor, or texture. Keeping some small animals (such as goldfish and hamsters) in the classroom will provide opportunities for students to learn new vocabulary related to major structures. Students should also learn that scientists are responsible for the ethical care of laboratory animals and that classroom animals deserve no less care (*Education Code* Section 51540).

STANDARD SET 3. Earth Sciences

Mountains, valleys, plains, rivers, lakes, and oceans are all features of the surface of Earth. Forces within Earth uplift the land; and the actions of wind, water, and ice carve Earth's surface into topographic features. Contrasts between rivers and oceans, mountains and deserts, and hills and valleys can become the natural settings for students to begin studying the earth sciences. Changing weather conditions (such as rain, wind, and temperature) provide students with opportunities to make observations and measurements. Recording changes in the weather provides a rich opportunity for class discussion and builds listening comprehension.

The materials that make up Earth's surface provide resources for human activities. Students learn that human consumption leads to waste that must be disposed of. This understanding will help them appreciate the importance of recycling and conserving Earth's resources.

> **3. Earth is composed of land, air, and water. As a basis for understanding this concept:**
>
> **a.** *Students know* characteristics of mountains, rivers, oceans, valleys, deserts, and local landforms.

Students can explore the variability of landforms by means of tangible experiences (such as making direct observations, hearing stories and seeing pictures, and making models on sand/water tables). They learn to identify the mountains, rivers, oceans, valleys, deserts, and other landforms in photographs or models. This activity will also help improve their vocabulary for describing things.

> **3. b.** *Students know* changes in weather occur from day to day and across seasons, affecting Earth and its inhabitants.

Students know that they do not wear the same clothes on a wet, windy day as they do on a hot, sunny day. They now need to extend their concept of the consequences of weather changes beyond their personal lives. Students make weather observations and note how the weather changes over a period of days, weeks, and months. They observe the generic effects of weather and seasons on the land and living organisms.

> **3. c.** *Students know* how to identify resources from Earth that are used in everyday life and understand that many resources can be conserved.

Students need to learn the connection between materials and the resources from which the materials were derived. Students learn the importance of science in understanding the need for good air to breathe and clean water to drink. Students may explore ways in which to conserve, recycle, and reuse materials, especially within the classroom and school site environment. It is important they learn that everything has an origin. For example, drinking water is derived from streams and lakes, wood and paper from trees, and bricks and metals from Earth.

STANDARD SET 4. Investigation and Experimentation

The ability to observe and describe common objects develops early and is enhanced by kindergarten instruction when students are introduced to the properties of solids and liquids, plants and animals, and landforms and weather conditions. Students can also be taught to compare and sort objects on the basis of the objects' properties and be encouraged to use mathematics to communicate some of their observations.

> **4.** Scientific progress is made by asking meaningful questions and conducting careful investigations. As a basis for understanding this concept and addressing the content in the other three strands, students should develop their own questions and perform investigations. Students will:
>
> a. Observe common objects by using the five senses. [*Caution:* Observational activities associated with tasting and smelling should be conducted only under parental supervision at home.]
>
> b. Describe the properties of common objects.
>
> c. Describe the relative position of objects using one reference (e.g., above or below).
>
> d. Compare and sort common objects by one physical attribute (e.g., color, shape, texture, size, weight).
>
> e. Communicate observations orally and through drawings.

Grade One — Science Content Standards

Students in grade one learn about the general properties of solids, liquids, and gases. They also learn about the needs of plants and animals and the functions of some of their external structures. Students also learn how to use simple weather-recording instruments, such as thermometers and wind vanes, and discuss daily and seasonal changes in weather. Students in grade one are adept at identifying the characteristics of objects and can either record those observations through pictures and numbers or begin to use written language. They can learn to make new observations when discrepancies exist between two descriptions of the same object or phenomenon.

The English–language arts standards require students to write brief expository descriptions about people, places, things, and events by using sensory details. Those expository descriptions may be aligned with the science standards that require students to record observations and data by using some written language. Teachers should guide students to respond to *who, what, when, where,* and *how* questions. Students expand their vocabulary by learning appropriate grade-level scientific terms (such as *freezing, melting, heating, dissolving,* and *evaporating*).

STANDARD SET 1. Physical Sciences

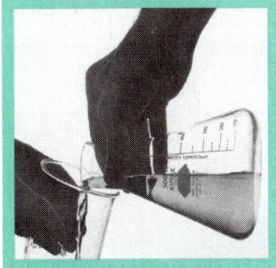

Students learn the general differences and similarities between properties common to all solids, liquids, and gases. The physical sciences standards in grade one provide a foundation for the study (in grade three) of evaporation and the changes in states of matter that may occur when solids and liquids are heated.

> **1. Materials come in different forms (states), including solids, liquids, and gases. As a basis for understanding this concept:**
>
> **a.** *Students know* solids, liquids, and gases have different properties.

Solids have definite shapes, meaning they are rigid and occupy a specific volume. This attribute distinguishes solids from liquids and gases whose fluid nature (or ability to flow) results in their shape being determined by the shape of whatever vessel contains them.

Teachers may demonstrate the properties of a solid by collecting a variety of solid objects of different shapes, sizes, weights, and textures. They demonstrate the fluid nature of a liquid by pouring water between same-sized measuring cups of different shapes. This demonstration shows that each cup holds the same amount of liquid even though the shapes are different. Distorting a partially inflated balloon into a variety of shapes shows that gases do not have a definite shape, and pushing the balloon into a container of water shows the amount or volume of water excluded by the gas.

Students can draw pictures and tell or write stories that illustrate the differences between the properties of solids, liquids, and gases.

> **1. b.** *Students know* the properties of substances can change when the substances are mixed, cooled, or heated.

Students can be taught that melting requires heating and freezing requires chilling. It may be helpful to use a thermometer to establish whether the temperature of a substance is increasing or decreasing before and after melting occurs. As ice water is heated, the temperature does not increase until the ice is melted. Students should begin to understand that some changes are reversible (e.g., ice melting) and some are irreversible (e.g., an egg cooking). Salt dissolved in water and recovered through evaporation may be cited as another example of a reversible process. Mixing baking soda with vinegar produces irreversible change, marked by the carbon dioxide gas bubbling up from the vinegar as the baking soda converts into soluble sodium acetate and water.

STANDARD SET 2. Life Sciences

Students in grade one are ready to focus on the favorable habitats (usually including air and soil), water, and energy supply (sunlight or food) that living organisms need to survive. Students will learn how plants and animals live in different environments and will discuss the relationship between structural form and function.

> **2. Plants and animals meet their needs in different ways. As a basis for understanding this concept:**
>
> a. *Students know* different plants and animals inhabit different kinds of environments and have external features that help them thrive in different kinds of places.

Students learn about the types of organisms that live in different environments and the ways in which they have adapted to their surroundings. Marine mammals off the Pacific coast typically have thick, blubbery skin (e.g., whales) or thick fur (e.g., sea otters) to withstand the cold water. Giraffes have long necks that help them to reach leaves near the tops of trees and spot predators from great distances. Those examples of adaptations are ones that students can readily discuss. Many stories and videos about plants and animals can help students learn about life on Earth.

> **2. b.** *Students know* both plants and animals need water, animals need food, and plants need light.

Learning what plants and animals need to survive is one of the foundations of ecology. Both plants and animals need water and air. Both also need a source of energy. Plants absorb sunlight, and animals eat food to meet their energy requirements. Plants and animals obtain what they need to survive through the environmental adaptations described above.

If plants are kept in the classroom, students can learn about their needs by caring for them. Students may enjoy field trips or walks in locations where shrubs and grasses attract small animals (particularly birds, lizards, and insects). To supplement those activities, the school library has many books with good stories about plants and animals that live in a variety of environments.

> **2. c.** *Students know* animals eat plants or other animals for food and may also use plants or even other animals for shelter and nesting.

This standard introduces students to the fact that all living organisms in an environment are interdependent. For example, some birds nest in shrubs and trees; insects (such as fleas) may inhabit dogs, cats, and other mammals. Animals may assist plant reproduction by spreading seeds.

Students can observe that insects eat the leaves of shrubs and grass and that this activity will attract additional small animals, such as birds and lizards, which eat the insects. Discussions of such observations introduce students to the idea of a food chain. Teachers should point out to their students that people are at the top of the food chain.

> **2. d.** *Students know* how to infer what animals eat from the shape of their teeth (e.g., sharp teeth: eats meat; flat teeth: eats plants).

This standard introduces the biological concepts of structural form and function, which are discussed extensively in later grades. A cat's sharp, pointed teeth are well suited for ripping and tearing the meat it eats, and the flat teeth of a cow are well suited to chewing and grinding the tough grasses it consumes. Students can study different specimens of teeth, including skeletal examples and fabricated models. They can find pictures of different kinds of teeth (carnivores and herbivores) in library books. Students can examine their own teeth by using mirrors and observe, record, and report to the class which teeth they use (front teeth or back teeth) when eating different types of food. The relationship between teeth and the food that animals eat may be taught during a field trip to the zoo or when a naturalist speaks to the students.

> **2. e.** *Students know* roots are associated with the intake of water and soil nutrients and green leaves are associated with making food from sunlight.

This standard is complementary to Standard 2.d as it emphasizes the relationship between plant structures and their functions. Students learn that roots take in water and nutrients from the soil. Green leaves are the sites where photosynthesis turns sunlight into food. If students have plants growing in the classroom, they may observe and record how the plants respond to different growing conditions. For example, a plant growing near a window may turn its leaves toward the light source and change its direction of growth to improve its ability to make food.

STANDARD SET 3. Earth Sciences

Students learn that each season has its own predictable range and trends of weather conditions. They also learn how to use simple equipment to measure weather conditions. To be prepared for studies in subsequent grade levels, students should learn that Earth receives energy from sunlight and that the warming of Earth has a strong influence on the weather.

> **3. Weather can be observed, measured, and described. As a basis for understanding this concept:**
>
> **a.** *Students know* how to use simple tools (e.g., thermometer, wind vane) to measure weather conditions and record changes from day to day and across the seasons.

Students learn how to use a thermometer and a wind vane to measure weather conditions. They may also make a simple rain gauge to improve the quality and detail of their weather observations, measurements, and records. In discussing their findings, they have opportunities to improve their vocabulary and expressive language. Students should have experience in recording day-to-day and seasonal changes in weather, but teachers should limit the time spent on those activities. For example, if students were to spend only ten minutes per day making measurements and discussing trends in the weather, the instructional time dedicated to this activity would amount to 30 hours over the course of a school year.

> **3. b.** *Students know* that the weather changes from day to day but that trends in temperature or rain (or snow) tend to be predictable during a season.

Teachers may wish to keep an eye on the weather report and allot instructional time to record weather conditions during a week in which precipitation or high winds are expected. They may also have students record data during a different week in which the weather is expected to be relatively stable. In bringing students' attention to those two differing sets of data, teachers may lead a spirited discussion. Although it is difficult to predict the weather, teachers should not encourage uninformed guesses. Historical data on temperature, wind, and rainfall conditions are typically collected for every city, often by stations located at airports. Those data are freely available on the Internet and are a useful resource.

> **3. c.** *Students know* the sun warms the land, air, and water.

Radiation from the Sun is ultimately responsible for atmospheric circulation and the weather, a fact that is introduced in grade one and mastered in grade five. Students in grade one may be made aware of the warming effect of the Sun's rays on their skin and may be shown that the air, land, and water are similarly warmed. For example, students can see that on a sunny day the asphalt of their playground is cool in the morning but hot by midday. On a cloudy day the asphalt may stay cool all day.

STANDARD SET 4. Investigation and Experimentation

Students continue to develop the ability to make quantitative observations and comparisons by recording and using numbers. Recording requires careful observing, comparing, and establishing the order of objects and events. Not to be overlooked is teaching students to revisit their observations. A revisit is best done when students find that they have different descriptions of the same object or event. Also important is the fact that an observation of change depends on having a "fixed" reference point. An object is known to have moved only because its position has changed in relation to a reference point.

> **4.** Scientific progress is made by asking meaningful questions and conducting careful investigations. As a basis for understanding this concept and addressing the content in the other three strands, students should develop their own questions and perform investigations. Students will:
>
> **a.** Draw pictures that portray some features of the thing being described.
>
> **b.** Record observations and data with pictures, numbers, or written statements.
>
> **c.** Record observations on a bar graph.
>
> **d.** Describe the relative position of objects by using two references (e.g., above and next to, below and left of).
>
> **e.** Make new observations when discrepancies exist between two descriptions of the same object or phenomenon.

Grade Two — Science Content Standards

In the physical sciences students in grade two learn about forces (pushes and pulls) and some common phenomena (such as gravity, magnetism, and sound). In the life sciences they learn about the life cycles of animals and plants and the basics of inheritance. Dogs always reproduce puppies, never kittens or hamsters; however, not all puppies look alike. There is both similarity within a species and natural variation, some of which is caused by the environment. In the earth sciences students learn that rocks are composed of different combinations of minerals, that smaller rocks and soil are made from the breakage and weathering of larger rocks, and that soils also contain organic materials. Students are introduced to fossils and the evidence they provide about Earth's history.

The content standards for both science and mathematics specify writing; measuring; simple graphing; and making drawings to record, organize, interpret, and display data. Students practice measuring (with appropriate tools) length, weight, temperature, and liquid volume, expressing those measurements in standard metric system units. Students in grade two should learn to organize their observations into a chronological sequence and be able to follow oral instructions for an investigation.

STANDARD SET 1. Physical Sciences

The primary aim of the physical sciences standards for students in grade two is to develop a foundation for the study of motion and force that will be developed still further at later grade levels. At a very basic level, students should learn about the forces of gravity and magnetism and the ability of vibrating objects to make sounds.

1. **The motion of objects can be observed and measured. As a basis for understanding this concept:**

    a. *Students know* the position of an object can be described by locating it in relation to another object or to the background.

Students learn how to locate an object by measuring its distance and noting its direction in relation to another object that serves as a reference point. It does not matter what the reference object is as long as it is stationary. Distances are measured by using metric system units (such as meters [m] and centimeters [cm]). Students also learn to describe the position of an object by its location in relation to a pattern in its background or to a very distant object.

> **1. b.** *Students know* an object's motion can be described by recording the change in position of the object over time.

This standard helps to develop concepts of motion (such as speed, velocity, and acceleration) that are formally taught in higher grades. Students in grade two may simply observe and record the position of objects at intervals of time and note the changes in speed and direction along the path the object travels. Some objects will travel farther than others in a fixed time and therefore travel faster.

> **1. c.** *Students know* the way to change how something is moving is by giving it a push or a pull. The size of the change is related to the strength, or the amount of force, of the push or pull.

Once set in motion, an object will move in a constant direction and at a constant speed unless it is pushed or pulled. A ball at rest will remain at rest unless it is given a push or a pull. Speeding up, slowing down, and changing direction all require a push or pull. The friction between surfaces that slows objects down may be considered a pull. Pushes and pulls may be related to corresponding changes in the position of an object and the speed or the direction in which it is traveling.

The term *force* is introduced into the vocabulary and will be used more formally in later grade levels. Students in grade two learn to think about pushes and pulls as forces and the strength of the push or pull as the strength or magnitude of the force. When students kick or throw a ball, the effort they put into the action of kicking or throwing is related to the force (push) that is applied to the object and therefore the speed and distance it will travel. They will also understand that the bigger the change in the object's motion, the bigger the push or pull that is required.

> **1. d.** *Students know* tools and machines are used to apply pushes and pulls (forces) to make things move.

Students become acquainted with some of the methods by which tools and machines transmit or apply pushes and pulls in order to make things move. The blow (push) of a hammer will move a nail into a piece of wood. A hit (push) by a baseball bat will change the direction of motion of a pitched ball. A car's engine applies a force to turn its wheels, causing the car to move. Students are taught that machines and tools allow people to apply and control forces, some of which may be much greater than they themselves could create. In grade three students will learn that a source of energy is needed for machines to do work and create the forces that move heavy objects or move objects at great speeds.

> **1. e.** *Students know* objects fall to the ground unless something holds them up.

Gravity is a fundamental attracting or pulling force. A table, chair, shelf, or bench can support an object by opposing this pull. From a student's perspective an object seems to push down on whatever supports it. By grade eight students will learn the supporting object is also pushing back. Students in grade two should be taught that all objects in the universe are pulled toward all other objects by a force called *gravity*. Bigger, more massive objects pull more strongly than lighter ones. When a child drops a toy, the toy falls toward the ground instead of floating up because the force of Earth's gravity pulls on the toy. (Of course, the gravity of the toy pulls on Earth equally as well.) Earth's gravitational force on an object is called the object's weight.

> **1. f.** *Students know* magnets can be used to make some objects move without being touched.

The poles of magnets either repel (push) or attract (pull) one another. The fact that magnetic pushes and pulls happen at a distance without direct contact should be carefully noted. Students may, for example, observe magnetic attraction and repulsion being transmitted through materials such as paper or tabletops. This concept will give students an understanding of force at a distance and of the ability to apply pushes and pulls to objects without touching them.

Many activities may be done with magnets. Teachers are encouraged to have bar and ring magnets with distinct north and south poles rather than refrigerator magnets made of alternating poles. In this way students can discover the north-south attraction and the north-north or south-south repulsion. Students may sprinkle iron filings on a piece of paper and observe what happens to the pattern that the filings make when a magnet moves under the paper. This experiment provides further evidence that magnets do not have to touch materials to attract or repel them.

> **1. g.** *Students know* sound is made by vibrating objects and can be described by its pitch and volume.

The primary objective is for students to know that vibrations (back and forth motions) produce sound. They should be able to describe both the pitch and volume of a sound and be able to distinguish between them. Teachers may allow students to touch various vibrating, sound-producing objects (such as tuning forks, drums, and stringed instruments) to demonstrate the relationship between an object's vibration and the sound it generates. Faster vibrations lead to higher pitches of sound. Students may enjoy hearing the tones produced by different musical instruments as they identify the vibrating sources of those sounds.

STANDARD SET 2. Life Sciences

Students learn that plants and animals have life cycles that are typical of their species. Students also begin to develop simple notions of inherited characteristics, variation within a species, and environmentally induced changes. Although the concepts are discussed at a simple level in grade two, they form a foundation for understanding the concepts of genetics, evolution, and ecology in later grade levels.

> **2. Plants and animals have predictable life cycles. As a basis for understanding this concept:**
>
> **a.** *Students know* that organisms reproduce offspring of their own kind and that the offspring resemble their parents and one another.

At one level the characteristics of a species are generally consistent from generation to generation. Dogs always give birth to puppies, and oak trees always drop acorns that grow into new oak trees. Offspring inherit genes from their biological parents with the result that they resemble their parents and each other. However, even among siblings, there is individual variation in both appearance and behavior. Some individual variation is due to genes that are inherited from each parent, as students will learn in later grade levels, and some is due to environmental influences.

> **2. b.** *Students know* the sequential stages of life cycles are different for different animals, such as butterflies, frogs, and mice.

The life cycles of some insects consist of egg, larval, pupal, and adult stages. Many organisms undergo molting processes during the larval stage or the adult stage. This phenomenon is typical of species that have tough external skeletons (e.g., grasshoppers, crabs). Using mealworms (obtainable from many pet stores and kept in plastic containers with bran meal) is a good way for students to watch the life cycle of grain beetles over a period of a few weeks. The life cycles of those insects can be slowed down by placing the containers in the refrigerator at night and over weekends. Mealworms molt as they grow during their larval stages, and the casings can be easily recovered for study. Frogs and many other amphibians also undergo a type of metamorphosis, but those changes unfold gradually. Mammals bear live young that resemble, to a great extent, their adult forms.

> **2. c.** *Students know* many characteristics of an organism are inherited from the parents. Some characteristics are caused or influenced by the environment.

This standard refines the understanding of Standard 2.a and prepares students for Standards 2.d and 2.e. As previously noted students must understand the concept of inheritance. Many characteristics of an individual organism are defined by the genes inherited from the biological parents. Other characteristics are strongly influenced by the environment or may be caused entirely by the environment.

> **2. d.** *Students know* there is variation among individuals of one kind within a population.

Offspring may generally look like their parents and each other but may still vary in such aspects as color, size, or behavior. Within a broader population the extent of variation may be greater still. This standard should be discussed in the context of the previous standard as variation is a function of both genetics and environmental influences.

> **2. e.** *Students know* light, gravity, touch, or environmental stress can affect the germination, growth, and development of plants.

It is relatively easy to change the appearance or growth patterns of plants by changing the conditions of growth. Discussion of this standard can be tied to the discussion of the previous two standards. Roots typically grow downward in response to gravity. Stems and leaves grow upward or sideways to seek sunlight or an artificial light source. Environmental stress resulting from inadequate light, lack of nutrients, or the wrong amount of water impedes or halts the growth of plants.

Students may learn about plant germination and growth by planting seeds and observing their development. The effects of environmental factors (e.g., the amount of water and light, types of soil) can be readily studied. Students can make a root change the direction of its growth by mounting the germinating seeds between wet paper towels, placing the towels on a vertical sheet, and turning the sheet after several days to make the root respond to the new perception of "down." This experiment is evidence that plant growth responds to the downward pull of gravity. Students can also study how plants grow toward light over a period of weeks and turn many of their leaves to face the sun over the course of a day. To observe the effects of environmental stress, students can vary the light, temperature, soil composition, and amount of water that groups of the same kind of seed receive. Students may have already studied and grown plants in grade one, so it is important that in grade two they study plants at a deeper level.

> **2. f.** *Students know* flowers and fruits are associated with reproduction in plants.

Plant germination and reproduction are related to the structure and function of seeds, flowers, and fruits. Although students may learn the idea entirely from stories and discussions, students can also take apart some large flowers to learn about plant reproductive structures. The use of magnifying glasses or simple microscopes will assist with observations. Inside the seed coat is the new plant embryo surrounded by food that is used during early growth and development. Students may find seeds inside juice oranges and apples and inside pea pods and ears of corn. Some common garden plants have extremely poisonous parts; therefore, the teacher should ensure that only safe plant materials are studied. Teachers should also be aware of student allergies.

STANDARD SET 3. Earth Sciences

The focus of earth sciences in grade two is on the composition, processes, and materials of Earth's crust. The term *weathering* is introduced as a process that leads to breaking rocks into smaller pieces. The interaction between the atmosphere and the upper surface of Earth's crust is the major source of weathering. Studying the relationship between weathering and soil formation, students learn that soil has an important effect on the growth and survival of plants. They also learn how soil is formed and about its constituent properties.

The concept of *geologic time* and the study of fossils are introduced. Students are asked to think abstractly about events that took place in Earth's ancient, geologic past. They will learn that Earth has not always looked the same as it does today. Teachers should present some of the evidence (particularly from fossils) that scientists use to "observe" what Earth was like in the geologic past. Students should be able to discuss and identify the origin of things they use in their everyday lives. Natural resources include rocks, minerals, water, plants, and soil. Understanding those ideas serves as a foundation for the study of earth sciences in later grade levels when students learn more about natural resources, including the identification of conservation techniques.

> 3. **Earth is made of materials that have distinct properties and provide resources for human activities. As the basis for understanding this concept:**
>   a. *Students know* how to compare the physical properties of different kinds of rocks and know that rock is composed of different combinations of minerals.

Students should know the physical properties (e.g., hardness, color, and luster) of a few of the most common minerals and be able to compare them. Students can compare rocks that are about the same size (volume) and note that some are heavier than others. They can also compare a few of the most common rocks. Students should conclude that rocks are composed of different combinations of minerals. They should know some simple techniques for making comparisons. It may be helpful to provide students with a set of common rocks formed from minerals (e.g., quartz, feldspar, mica, hornblende). Coarse-grained rocks (e.g., granite, gabbro, diorite) allow students to see individual mineral grains. The use of magnifying glasses or simple microscopes can help students to sort and classify the rocks according to their constituent minerals.

> **3. b.** *Students know* smaller rocks come from the breakage and weathering of larger rocks.

Through the process of weathering (interaction between the atmosphere and Earth's surface), large rocks break down into smaller rocks. Rocks and minerals reduced by weathering to a very small size eventually turn into soil. Weathering may be a physical or chemical process. Physical weathering occurs when big rocks break down from the repeated freezing and thawing of water in cracks or when rocks are wedged apart by root growth. Rocks may also be chemically weathered through reactions with constituents of the atmosphere.

> **3. c.** *Students know* that soil is made partly from weathered rock and partly from organic materials and that soils differ in their color, texture, capacity to retain water, and ability to support the growth of many kinds of plants.

This standard looks at soil as a whole and calls upon students to examine organic soil constituents in addition to weathered rock. Various combinations of weathered rock and organic material are reflected in soil properties, such as color, texture, capacity to retain water, and fertility. Organic materials, such as rotting, dead leaves and twigs as well as animal remains, add to the results of weathering. Burrowing mammals, such as gophers, and worm activity are responsible for mixing the soil. The types of both weathered material and organic remains that are mixed together and the proportions of the mixed constituents affect the properties of the soil. Dark soils often contain organic material, and red soils often derive from rocks and minerals rich in iron. Soil fertility depends more on the organic material than on the weathered rock contained in the soil. Decaying organic materials act to hold moisture in a spongelike manner and return nutrients to the soil.

> **3. d.** *Students know* that fossils provide evidence about the plants and animals that lived long ago and that scientists learn about the past history of Earth by studying fossils.

A fossil is a physical record of life that lived in the geologic past. The study of fossils provides an opportunity for students to investigate plants and animals that lived long ago. Scientists compare plant impressions and the footprints and skeletons of dinosaurs to the characteristics of modern plants and animals. This study yields clues about the environments in which ancient organisms once lived. Geologists apply the concept of *uniformitarianism* as they try to reconstruct ancient geologic environments. The present is a key to the past if one assumes slow, uniform, and sequential changes have led to present conditions.

Teachers may be able to obtain a small collection of fossils or manufactured copies of fossils from local libraries, museums, science centers, or universities and supplement the collection with pictures of fossils found in books. By using fossils or pictures of fossils, students can try to reconstruct what animals and plants might have looked like when alive. For example, fossil leaves are often black, but were they black when they were alive? Students may report their results orally or in written form, presenting their drawings as support for their conclusions.

> **3. e.** *Students know* rock, water, plants, and soil provide many resources, including food, fuel, and building materials, that humans use.

Resources to meet many human needs, such as food, clothing, fuel, and shelter, originate from rocks, water, plants, and soil. Students should understand the relationship between manufactured materials and the natural resources from which they originate. For example, humans use the same weathered rocks that serve as a source material for soil as a resource for manufacturing building materials. Soil supports plant growth, and plants supply food for humans and for some of the animals that humans eat. Plants also supply fuel and building materials. Students should be able to name and identify the origin of the resources of some of the things they use as food, clothing, and shelter.

STANDARD SET 4. Investigation and Experimentation

The power of science is its ability to predict what will happen on the basis of concepts, principles, and theories related to the natural world. In grade two students can observe patterns associated with changes in objects and events. Under similar conditions students can use these patterns to make simple predictions. Teachers should not confuse predictions with hypotheses, which will be introduced in grade six. Measurements in science are always associated with units, and this idea is an important lesson for students to learn as they measure and record. Students should learn to measure length in meters and centimeters, weight (mass) in grams and kilograms, volume in liters and milliliters, and temperature in degrees Celsius. Students should also learn that scientists use tools to extend their powers of observation. Simple magnifiers and microscopes can be used to reveal exciting and sometimes surprising microstructures and properties of common objects, such as sand and cloth. Making careful sketches of what is observed under the magnifier is an important method by which students can communicate their observations.

> **4.** Scientific progress is made by asking meaningful questions and conducting careful investigations. As a basis for understanding this concept and addressing the content in the other three strands, students should develop their own questions and perform investigations. Students will:
>
> **a.** Make predictions based on observed patterns and not random guessing.

Chapter 3
The Science Content Standards for Kindergarten Through Grade Five

Grade Two

b. Measure length, weight, temperature, and liquid volume with appropriate tools and express those measurements in standard metric system units.

c. Compare and sort common objects according to two or more physical attributes (e.g., color, shape, texture, size, weight).

d. Write or draw descriptions of a sequence of steps, events, and observations.

e. Construct bar graphs to record data, using appropriately labeled axes.

f. Use magnifiers or microscopes to observe and draw descriptions of small objects or small features of objects.

g. Follow oral instructions for a scientific investigation.

Grade Three — Science Content Standards

Students in grade three are introduced to some of the most fundamental patterns in nature and should be taught that science makes the world understandable. For example, by observing that the stars appear fixed in relation to one another, one can identify five planets in motion against the starry background. Students in grade three begin to build a foundation for understanding the structure of matter and forces of interaction. They will study the properties of light and gain an appreciation for how light affects the perception of direction, shadow, and color. Students in grade three will also extend their knowledge of ecology by learning about different environments, such as oceans, deserts, tundra, forests, grasslands, and wetlands, and the types of organisms adapted to live in each.

The curriculum and instruction offered in grade three enable students to read materials independently with literal and inferential comprehension and to support answers to questions about the material by drawing on background knowledge and details from the text. Instruction in information literacy that incorporates library resources will help students become skilled in locating information in texts by using titles, tables of contents, chapter headings, glossaries, and indexes. The science standards complement the mathematics standards by asking students to predict future events on the basis of observed patterns and not by random guessing.

STANDARD SET 1. Physical Sciences (Energy and Matter)

The discussion of energy and matter in grade three is at a simple level, but it sets a foundation for further study in later grade levels. Students learn that energy may be stored in various ways and that both living organisms and machines convert stored energy into heat and motion. Matter will also be studied in more detail than at the previous grade levels. Atoms will be introduced as the smallest component of the elements that compose all matter. Students will learn that there are different kinds of atoms and that their names and symbols are displayed on the periodic table of the elements. This standard set will prepare the students for a more detailed treatment of the properties of the elements and their combinations in grade five.

1. **Energy and matter have multiple forms and can be changed from one form to another. As a basis for understanding this concept:**

    a. *Students know* energy comes from the Sun to Earth in the form of light.

Energy is a physical attribute capable of causing changes in material objects. This concept is one of the more important ones in science. At a simple level, and certainly for the treatment of the subject in grade three, energy is the ability to do work; to make things move, stretch, or grow; or to cause physical and chemical changes. Throughout the study of science, many more forms of energy and their effects will become evident. Students in grade three should understand that Earth's major source of energy is the Sun and that the Sun's energy is seen as light and felt as heat. It is important for students to realize that although light and heat are not exactly the same, both are forms of energy.

> **1. b.** *Students know* sources of stored energy take many forms, such as food, fuel, and batteries.

Students should understand that the energy stored in food, fuel, and batteries can be released to create useful motion, light, and heat. For example, students may study the components of a flashlight and leave it on until the light goes out to emphasize that batteries store a limited amount of energy. Matches and candles are cold before lighting; but when burned, their stored energy is released in the form of light and heat. Students should be taught that they eat food in order to use its stored energy to make it possible for them to grow, maintain their warm body temperature, and be able to work and play. Teachers should note that all those forms of stored energy are contained in chemical substances and released through chemical changes.

> **1. c.** *Students know* machines and living things convert stored energy to motion and heat.

This standard expands concepts introduced in earlier grades. The way in which machines and living things take different sources of energy and produce useful heat and motion should be examined in greater detail. An automobile engine releases the chemical energy stored in gasoline (and air) and uses it to turn the wheels and move the vehicle. Some students may be familiar with wind-up toys and will be able to understand that the potential energy stored in springs is used to turn the gears that activate the toy. Similarly, the energy stored in natural gas is converted to heat in a gas stove, oven, or furnace. Students learn that food is broken down into smaller components; some components are carried to the muscles, where the energy stored is released as movement and as heat, keeping the human body warm.

> **1. d.** *Students know* energy can be carried from one place to another by waves, such as water waves and sound waves, by electric current, and by moving objects.

Energy movement or transfer should be discussed in terms of moving objects (e.g., thrown balls), waves (e.g., light, sound, seismic or earthquake waves, and ocean waves), and electricity (charges passing through a wire). The key point in this standard is that energy is carried in those forms and transferred from one place to another. Simple toys that demonstrate transfer of motion to another object are good examples of this principle and form the foundation for understanding the conservation of energy. Energy of motion is transferred into heat through friction (such as when students rub their hands together rapidly and feel the heat generated by the rubbing motion).

Students can also study how waves transfer energy from one place to another through a medium (water or air), with no net motion or flow of matter. Students can demonstrate this principle by creating waves in a tub of water that contains materials (e.g., cork stoppers or small balls) floating on the water. Energy is required to start the wave at one end and is then transferred to the objects in the water, generating a bobbing motion. The students should note that this action can be accomplished without any net transfer of water from one end to the other. They should observe that waves make floating objects bob up and down and back and forth, but the objects stay in essentially the same position as they were in before the waves were generated.

Sound is made by vibrating objects and is carried in compression waves through the air. Sound can create vibrations in a distant second object (such as an eardrum) without direct physical contact between the two objects.

The evidence for electrical energy transfer surrounds students in their everyday lives. Electrical energy comes from power plants that may use fossil fuels, water, wind, or nuclear power. The key idea is that electrical energy has a source, is carried in wires as electricity, and is converted to more easily recognized forms of energy (such as heat, light, and motion).

> **1. e.** *Students know* matter has three forms: solid, liquid, and gas.

Students in grade three must understand that matter is a substance that occupies space and may assume the form of a solid, liquid, or gas. Students should view pictures and read articles about lava and molten steel to make the point that most substances can turn to liquid when heated to a high enough temperature. Likewise, a gas can turn to a solid if sufficiently cooled. For example, carbon dioxide, a gas at room temperature, can be frozen into dry ice.

> **1. f.** *Students know* evaporation and melting are changes that occur when the objects are heated.

This standard is an extension of what students will have learned about water in kindergarten and grade one. New to them is the generalization that melting and evaporation are processes that may occur when substances other than water are heated. Books and videos from the school library that show the process of making iron and steel may be helpful in providing instruction on this standard.

Chapter 3
The Science Content Standards for Kindergarten Through Grade Five

Grade Three

> **1. g.** *Students know* that when two or more substances are combined, a new substance may be formed with properties that are different from those of the original materials.

This standard introduces the idea that pure substances have a fundamental character that is necessary in order to distinguish chemical changes from physical changes. Students are asked to build on some concepts that were introduced in earlier grades concerning changes in state and properties that may occur when two substances are mixed and react. Some students may begin to realize that there is a difference between mixtures and pure substances. The focus is on the new and different properties that are formed when two or more substances are mixed. The chemical reaction that occurs between baking soda and vinegar, producing carbon dioxide (and sodium acetate and water), is one of several simple reactions that may be used to illustrate this difference. Teachers can also use the burning of a candle to demonstrate this concept. The products with very different properties are the carbon dioxide gas and the soot produced and the heat and light released. Water vapor, also formed by burning a candle, may be observed as condensation on a cool object held above the candle.

> **1. h.** *Students know* all matter is made of small particles called atoms, too small to see with the naked eye.

The important idea to convey is that all familiar substances are made of *atoms,* the term for the smallest particles of matter that retain the properties of the elements. To understand atoms, students must first be introduced to the idea that *matter* is the general name given to anything that has mass and occupies space. They should then be taught that matter comprises all solids, liquids, and even invisible gases. Just as a brick wall consists of many individual bricks, all matter consists of smaller bits that combine to make up what is seen. Students can discover this principle by looking through an inexpensive 30-power (30x) microscope to discover that the apparently solid colors on the cover of magazines actually consist of repeated patterns of colored dots.

Atoms are so tiny that detection requires techniques that go beyond the power of conventional microscopes. The following imaginary experiment may be helpful in understanding the basic concept of the atom. If a student were to take an object made of a pure element, such as a piece of aluminum foil, and cut it in half, both halves are still aluminum. If each of these pieces is then cut in half a second, third, fourth, and fifth time, the pieces become progressively smaller but are still aluminum. Is it possible to keep cutting the pieces in half forever and still have a piece of aluminum? How small must a piece be so that at the next cut it will no longer be aluminum? In pondering this question, early philosophers concluded that there must be a very small but indivisible piece of matter that still has the properties of aluminum or any other element. They named these smallest pieces *atoms.*

1. i. *Students know* people once thought that earth, wind, fire, and water were the basic elements that made up all matter. Science experiments show that there are more than 100 different types of atoms, which are presented on the periodic table of the elements.

In ancient times people believed that everything was made of combinations of just four elements: earth, air, fire, and water. This belief is understandable when one observes a log burn to become ash, fire, and hot gases, some of which condense into water. The Greeks, however, conjectured that matter is made of tiny particles. Today this belief is known to be true, and those particles are called *atoms.* More than 100 different types of elements are displayed on the periodic table of the elements. Students in grade three should know a chart exists that displays the names and symbols of known elements and other information.

The names of elements may fascinate students. Many elements may be familiar to the students (e.g., gold, silver, copper, iron, oxygen), but some will not be familiar. Students may enjoy finding names of familiar elements on the table. The custom of science is that discoverers have the right to name their elements. Some elements, such as einsteinium and seaborgium, are named after famous scientists (Albert Einstein and Glenn Seaborg) whose personal lives are the basis of interesting stories. Students may be fascinated to learn that one element is named californium and one is named for the city of Berkeley: berkelium. An important concept for students to know is that any substance not listed on the periodic table comprises a combination of different types of atoms (elements) that are listed. Living organisms, for example, are mostly made up of carbon, oxygen, nitrogen, and hydrogen atoms.

STANDARD SET 2. Physical Sciences (Light)

Light, like heat, is a form of energy. Standard Set 2 calls for students to know some of the properties of light but does not require them to understand light as energy in a waveform. They should know that light travels in a straight line away from its source and that the color of an object is affected by the color of light that strikes it.

2. Light has a source and travels in a direction. As a basis for understanding this concept:

  a. *Students know* sunlight can be blocked to create shadows.

Teachers may draw an analogy between an opaque object casting a shadow in sunlight and the dry place created when an umbrella blocks the fall of raindrops. The energy of sunlight is absorbed by the opaque object and is prevented from passing through to the ground. Students should be encouraged to experiment with shadows and to think about the source and direction of the light. They can cut cardboard into different shapes, compare the size and shape of the cardboard with the size and shape of its shadow, and notice whether the edges of the shadows are sharp or fuzzy.

> **2. b.** *Students know* light is reflected from mirrors and other surfaces.

Light reflected from a mirror or other surface changes direction by reflection and then continues to travel in a straight line. To demonstrate reflection and note the path of a reflected beam of light, the teacher or a group of students can use chalk dust or a water mist to help trace the path of the light beam in a darkened room.

> **2. c.** *Students know* the color of light striking an object affects the way the object is seen.

Two factors determine the color of an object: the color of the light illuminating the object and the interaction of the light with the object; for example, which colors are absorbed and which are reflected. Students can see this principle for themselves by being asked to describe an object's color viewed under lights of different colors. Because sunlight contains all the colors of the rainbow, light sources of different colors can be created by passing sunlight through colored cellophane. This principle can also be demonstrated by using colored light bulbs. To explore the principle in the standard, students may observe that a white object will be seen as the color of the light that illuminates it. For example, if a white piece of paper is seen under red lights, it appears to be a red piece of paper.

> **2. d.** *Students know* an object is seen when light traveling from the object enters the eye.

Light is a form of energy to which the eye is sensitive. An object can be seen because the light that travels from the object enters and interacts with the eye. If opaque material comes between the eye and an object being viewed, the opaque material blocks the light and the object disappears from view, demonstrating that light travels in a straight line.

STANDARD SET 3. Life Sciences

The life sciences standards in grade three continue to develop students' concepts of ecology and evolution by relating adaptation to the survival and fitness of the organism. Although natural selection is not formally discussed at this grade level, the foundation is set for teaching that principle in later grade levels. A significant effort is made to enhance students' knowledge of the types of plants and animals in different environments as this understanding becomes an important base of knowledge. These standards challenge students to consider the effects of environmental changes on organisms. The concept of extinction is introduced, and organisms in the fossil record are compared to contemporary organisms.

> **3. Adaptations in physical structure or behavior may improve an organism's chance for survival. As a basis for understanding this concept:**
>
> **a.** *Students know* plants and animals have structures that serve different functions in growth, survival, and reproduction.

Students have learned about the roots and leaves of plants in grade one and the functions of flowers and fruit in grade two. Many other external structures of plants and animals (e.g., cactus thorn, porcupine quill, crab shell, bear claw, and kangaroo pouch) serve important functions, and students in grade three will recognize many common examples through reading and observing examples from nature. This standard can be taught in the context of the one that follows and can serve as the basis for extended study and discussion.

> **3. b.** *Students know* examples of diverse life forms in different environments, such as oceans, deserts, tundra, forests, grasslands, and wetlands.

The organisms that live in oceans, deserts, tundras, forests, grasslands, and wetlands are different from one another because their environments are different. For example, animals with thick fur are able to survive a cold habitat. Gills allow fish to obtain oxygen from water, whereas lungs allow mammals to obtain oxygen from the atmosphere. Desert plants and animals have adapted by conserving the small amount of water they require. The thick, waxy leaves of some plants prevent water loss. Many desert animals are nocturnal and search for food during the cool of night.

Students should be taught about Earth's different habitats or *biomes* and be able to describe the characteristics of some of the plants and animals living in each. Students should be encouraged to locate information in nonfiction books and other library resources and be able to describe how living organisms are adapted for survival in their particular biome.

> **3. c.** *Students know* living things cause changes in the environment in which they live: some of these changes are detrimental to the organism or other organisms, and some are beneficial.

Living organisms, including humans, inevitably cause changes (some minor and some major) in the environment as the organisms compete for food, shelter, light, and water. Those changes are different from external changes, such as a fire started by lightning or flooding related to excessive rainfall. When some organisms become more or less successful in their quest for survival, the environmental balance changes and so does the environment. For example, beavers build dams that block streams, forming small lakes in which they can then reside. This activity is

Chapter 3
The Science Content Standards for Kindergarten Through Grade Five

Grade Three

beneficial to plants and animals that prefer to live in still water, but it is detrimental to plants and animals that are used to living in an open stream. It is also detrimental to large trees as they are cut and become material for the beaver. Trees affect the environment by blocking the sunlight; consequently, the felling of a large tree by a beaver may benefit smaller plants and shrubs that can now grow in its place. These examples are some of the types of environmental relationships studied by ecologists.

> **3. d.** *Students know* when the environment changes, some plants and animals survive and reproduce; others die or move to new locations.

Many plants and animals have specialized structures that allow them to survive and reproduce in the environment in which they live. Consequently, they may be adversely affected by environmental changes. For example, many plants and animals not suited to desert conditions will die if their environment becomes dry and desertlike for an extended period of time. Plants and animals establish a balance with one another in their shared environment. Consequently, environmental changes that affect one or more plants or animals in a given biome may eventually affect all living organisms in that biome. Animals may move, and seeds may be blown or carried to new, more favorable locations.

> **3. e.** *Students know* that some kinds of organisms that once lived on Earth have completely disappeared and that some of those resembled others that are alive today.

When an environment changes more quickly than a species of animal or plant can adapt, that species may become extinct. Fossils provide numerous examples of extinct plants and animals. By studying the characteristics of fossils, students can see that some extinct animals resemble animals that are alive today and that others are quite different.

Students can relate modern animal remains to the environments from which they came and then they can apply the same types of observations and reasoning to determine the kind of environment that may have supported the fossilized animals and plants.

STANDARD SET 4. Earth Sciences

Earth sciences standards in grade three center on the concept that objects in the sky move in regular and predictable patterns. It is important that students know and are familiar with the patterns and movements of the Sun, Moon, and stars, both as those bodies actually move and as they appear to move when viewed from Earth. Seasonal changes correlate with changes in both the amount of daily sunlight and the position of the Sun in the sky. Seasonal changes are caused by the tilt of Earth's axis of rotation and the position of Earth relative to the Sun. Students will also learn about the relationships between the phases of the Moon and the changes in the positions of the Sun and Moon. Using models and telescopes may help students grasp the concepts presented in the standards.

> **4. Objects in the sky move in regular and predictable patterns. As a basis for understanding this concept:**
>
> **a.** *Students know* the patterns of stars stay the same, although they appear to move across the sky nightly, and different stars can be seen in different seasons.

The relative position of stars with respect to each other in the night sky is fixed. The apparent motion of the stars through the night sky is a function of Earth turning on its own axis. Starlike objects do move across the fixed pattern of stars in the night sky, but those "stars" are really planets. Stars appear stationary relative to one another because they are far outside the solar system. The positions of stars appear to change each season from a particular point of view on Earth because that point will face progressively different parts of the universe at night. The stars that are visible in the summer nighttime sky would be visible in the winter daytime sky if they were bright enough to outshine the Sun.

> **4. b.** *Students know* the way in which the Moon's appearance changes during the four-week lunar cycle.

Students should be taught to observe the phases of the Moon; recognize the pattern of changes; and know such terms as the *full, quarter, waxing, waning,* and *crescent Moon.* The reason for this pattern of changes may then be explored.

One side of the Moon is always in sunlight (except in the case of an eclipse). How much of the sunlit surface of the Moon will be visible from Earth depends on the relative positions of Earth, the Moon, and the Sun. Earth and the Moon continuously cycle through changes in their positions relative to the Sun; therefore, the Moon will go through phases from "new" to "full" depending on how much of its lighted surface is visible from Earth.

Models may help in the teaching of the standard. Students may be shown the rotation of Earth on its axis; how the day and night cycle works; and why the Moon, like the Sun, appears to rise and set. Students may also be shown Earth's position relative to the Sun, the Moon's position relative to Earth, and how Earth orbits the Sun once a year. Students can observe the actual position changes in the Moon and in the background star patterns at the same time each night, continuing their observations long enough to include a full lunar cycle. They can be shown how the motion of the Moon around Earth accounts for those observations.

Chapter 3
The Science Content Standards for Kindergarten Through Grade Five

Grade Three

> **4. c.** *Students know* telescopes magnify the appearance of some distant objects in the sky, including the Moon and the planets. The number of stars that can be seen through telescopes is dramatically greater than the number that can be seen by the unaided eye.

Students are often startled the first time they look at details of the Moon through a telescope or even through high-quality binoculars. They quickly come to appreciate how those instruments facilitate the study of very distant objects. With the help of a telescope or very high-powered binoculars, students can see the rings of Saturn and some of the details of other planets. Students must never be permitted to look directly or stare at the Sun with the naked eye through binoculars, telescopes, or any other optical instruments. There are many pictures taken by powerful telescopes of planets, stars, and galaxies that students should have the opportunity to study in books.

> **4. d.** *Students know* that Earth is one of several planets that orbit the Sun and that the Moon orbits Earth.

The patterns of the stars stay the same relative to one another although they appear to move because of the rotation of Earth. Several starlike objects move across the sky's star patterns. They are planets that shine by light reflected from the Sun. Five planets can be seen without the aid of a telescope: Mercury, Venus, Mars, Jupiter, and Saturn. Three can be seen only with the aid of a telescope: Uranus, Neptune, and Pluto. Earth is also a planet and moves about the Sun in a path (orbit) that is similar to that of the other planets. Nine planets are in the solar system.* The Moon orbits Earth. Because Earth itself is a planet, measuring the orbits of other planets is a complex process. The process is so complex that scientists took a long time to figure out the different spatial relationships between the Moon, Earth, other planets, and the Sun.

> **4. e.** *Students know* the position of the Sun in the sky changes during the course of the day and from season to season.

During a single day the rotation of Earth causes the position of the Sun to change on the horizon. It may be helpful for students to keep track of the Sun's position and watch how shadows lengthen rapidly as sunset approaches. From season to season the length of day and the angle of the Sun vary. Students should know that they live in the Northern Hemisphere, where the Sun at noon is lower and to the south in the sky in the winter and more directly overhead in the summer. Shorter or longer days and more or less direct sunlight characterize the seasons. The angle of the Sun in the sky at noon and the length of the day vary throughout the year because Earth's axis is tilted in comparison to the plane of its orbit.

*Under resolutions passed by the International Astronomical Union on August 24, 2006, there are eight planets. Pluto no longer meets the definition of a "planet" but is now classified under a new distinct class of objects called "dwarf planets."

STANDARD SET 5. Investigation and Experimentation

Children should be taught to make careful measurements, but they also need to learn that some errors in measurement are unavoidable. Sometimes errors arise through carelessness, misuse of measurement instruments, or recording mistakes. These human errors can be minimized by instruction and practice in measuring carefully and properly and by double (or triple) checking of measurements. Even then errors may be introduced because of limitations in the precision of the instruments used to make the measurements. Students should be taught how to make the most precise measurements possible with the tools available. They should also repeat their measurements several times. Sometimes they will obtain results that are different each time. If those differences are significant, students should examine their measurement methods to see whether an obvious error occurred.

Students can begin to make predictions based on observations, prior knowledge, and logic. Predictions should not be confused with random guesses. Students should know that their predictions must be verified by experiments and the analysis of data gathered from careful measurements.

5. Scientific progress is made by asking meaningful questions and conducting careful investigations. As a basis for understanding this concept and addressing the content in the other three strands, students should develop their own questions and perform investigations. Students will:

  a. Repeat observations to improve accuracy and know that the results of similar scientific investigations seldom turn out exactly the same because of differences in the things being investigated, methods being used, or uncertainty in the observation.

  b. Differentiate evidence from opinion and know that scientists do not rely on claims or conclusions unless they are backed by observations that can be confirmed.

  c. Use numerical data in describing and comparing objects, events, and measurements.

  d. Predict the outcome of a simple investigation and compare the result with the prediction.

  e. Collect data in an investigation and analyze those data to develop a logical conclusion.

Grade Four — Science Content Standards

Students in grade four will learn to design and build simple electrical circuits and experiment with components such as wires, batteries, and bulbs. They will learn how to make a simple electromagnet and how electromagnets work in simple devices. They will observe that electrically charged objects may either attract or repel one another and that electrical energy can be converted into heat, light, and motion. Students in grade four expand their knowledge of food chains and food webs to include not only the producers and consumers they have previously discussed but also the decomposers of plant and animal remains, such as insects, fungi, and bacteria. They will also learn about other ecological relationships, such as animals using plants for shelter or nesting and plants using animals for pollination and seed dispersal. Students in grade four study rocks, minerals, and the processes of erosion. They also study the processes of weathering and erosion as a way of leading into the study of the formation of sedimentary rocks.

Students in grade four learn to formulate and justify predictions based on cause-and-effect relationships, differentiate observation from inference, and conduct multiple trials to test their predictions. In collecting data during investigative activities, they learn to follow a written set of instructions and continue to build their skills in expressing measurements in metric system units. They will analyze problems by identifying relationships, distinguishing relevant from irrelevant information, sequencing and prioritizing information, and observing patterns, all of which support the *Mathematics Content Standards*.[2] They should conduct scientific investigations and communicate their findings in writing.

STANDARD SET 1. Physical Sciences

Students entering grade four have already had some exposure to the subjects of electricity and magnetism, but these standards are a systematic effort to develop the principles of each and show how they are interrelated. The standards in grade four provide a simple understanding of electricity and magnetism and some applications in everyday life; they help to develop a foundation for further learning in high school.

1. **Electricity and magnetism are related effects that have many useful applications in everyday life. As a basis for understanding this concept:**
    a. *Students know* how to design and build simple series and parallel circuits by using components such as wires, batteries, and bulbs.

Students should design and build series and parallel circuits with wires, batteries, and bulbs. Many science books describe simple experiments for constructing series and parallel circuits. In series circuits one wire loop connects all the components, so current flows sequentially through the components in the one loop. In parallel circuits several loops of wires connect the components. A simple series circuit consists of two or three light bulbs wired together with a battery in a single loop. If the filament of one bulb breaks (or one bulb is removed from its socket), the single-circuit loop is broken and all the lights go out. Teachers may make a parallel circuit by extending two wires, parallel to each other, from the poles of a battery. Then they connect two or three bulbs, individually, across the parallel wires. If one of the bulb filaments is broken, the other bulbs still remain lit.

The series circuit is like a circular road that has no intersection; a series circuit has only a single path, and all components must carry the same current. The amount of current that can flow through a circuit depends on resistance. The lower the circuit's resistance, the higher the current that can flow through it. Overall resistance in a series circuit is the sum of the resistances of its individual components. In parallel circuits there are intersections and alternate pathways for the current, and each pathway may have different components on it. These alternate pathways split the current between them, depending on their electrical resistance (again, lower resistance along a pathway allows higher current). An alternate pathway with extremely low resistance, such as a wire with no components on it, is sometimes called a *short circuit*. Short circuits can prevent the rest of the circuit from operating properly and be dangerous because the short-circuiting wire may become very hot.

> **1. b.** *Students know* how to build a simple compass and use it to detect magnetic effects, including Earth's magnetic field.

Students should know that all magnets have two poles: north and south. They should have already experienced the attraction of the north and south poles and the repulsion of north-to-north and south-to-south in their science studies in grade two. They should have noted that the repulsion or attraction is stronger when the poles are close and weaker when the poles are further from each other. Any magnet suspended so that it can turn freely will align with Earth's magnetic field, provided that the attraction is not overwhelmed by a stronger local field. A compass needle will detect and respond to the presence of magnets.

Students can build a simple compass by rubbing a craft needle (blunt-tip) on a strong permanent magnet to magnetize it. (*Caution:* Students should be closely supervised to avoid injury.) The magnetized needle may then be placed on a piece of cork or sponge floating in a small bowl of water. The same effect may be observed by using a small (one centimeter long) bar magnet resting on a floating piece of cork. The magnet or needle will float around until one end points generally toward Earth's magnetic north pole (which is slightly off from true north). The north end (pole) of a magnet refers to its attraction to Earth's magnetic north pole. Ring magnets may be suspended from a thread, and the axis of the hole in the magnet will also point north and south. If the students use a commercial compass, they can confirm the orientation of their experimental compass. When working with compasses, teachers need to check that a set of compasses all point in the same direction.

Large amounts of steel (such as the supporting beams of modern buildings) and operating electronic devices (such as television sets and computers) may distort the effects of Earth's magnetic field and cause an inaccurate compass reading.

> **1. c.** *Students know* electric currents produce magnetic fields and know how to build a simple electromagnet.

Once students understand that electric currents produce magnetic fields, they can apply this knowledge to the construction of a simple electromagnet made by wrapping a half-meter length of insulated wire around a large iron nail or iron rod and connecting the ends of the wire to a battery. When an electric current from the battery flows through the wire, the iron bar is magnetized. Students can use a compass to prove that, like permanent magnets, their electromagnet has two poles. If the orientation of the battery is reversed, the poles of the magnet are reversed so that south becomes north and north becomes south. Students in grade four are unlikely to have sufficient knowledge to predict that this would happen, but in high school they will have an opportunity to understand the principles at a much deeper level.

> **1. d.** *Students know* the role of electromagnets in the construction of electric motors, electric generators, and simple devices, such as doorbells and earphones.

This standard builds on the previous one by challenging students to become aware of the role of electromagnets in their surroundings at home and at school. They learn that electromagnets are important in the function of electric motors, generators, doorbells, and earphones. Electromagnets may be thought of as magnets that can be turned on and off. When an electromagnet is switched on, an adjacent iron magnet can be made to move; this type of electrically induced movement can be harnessed (e.g., to ring doorbells or vibrate a speaker element in a pair of headphones). Simple schemes for constructing the equivalent of a doorbell or an electric motor may be found in textbooks. Constructing such a device helps students realize that they are using the interaction of two magnetic fields: one from a coil (an electromagnet) and the other from a permanent magnet (or, in some schemes, another electromagnet).

One way to understand a motor is that the alternating attraction and repulsion of the two magnetic fields converts electrical energy into the energy of motion. In high school, students will learn that charges (electrons) flowing in wires that cross magnetic fields experience a force that explains the rotation in the motor. An electric generator acts like an electric motor operating in reverse to change the energy of motion into electrical energy.

To go further in understanding the workings of home appliances, students may consult books or multimedia references on how things work. Students should be warned not to dismantle electrical appliances at home. *Note:* Dismantling electrical devices to determine how they work may be dangerous because some devices can hold lethal charges for days or weeks.

> **1. e.** *Students know* electrically charged objects attract or repel each other.

After scuffing their feet on a carpet on a dry day, students may have had the experience of "getting a shock" from touching a grounded object or another person. This experience is an example of *static electricity,* which is associated with the gain or loss of negative electric charges (electrons). The shock a student receives in the foregoing example is associated with equalizing the charges between objects and involves the movement of electrons. Attractive and repulsive forces are at play between charged objects; however, lightweight objects (such as balloons and scraps of paper) are typically needed to perceive these forces at work.

A negatively charged object will attract a positively charged object and repel another negatively charged one. Two positively charged objects will also repel each other. This phenomenon is analogous to the like poles of magnets repelling each other and the unlike poles attracting. A latex balloon may be used to demonstrate attraction. The balloon may be suspended from the ceiling by a thread and rubbed with a wool sock. The balloon and sock attract each other. The teacher may rub with a sock another balloon (attached to a stick by a 20- to 30-centimeter-long thread) next to the suspended balloon. The two balloons will repel each other. Balloons that have been "charged" in this way may also pick up little bits of paper (1 mm square) and attract strands of hair.

> **1. f.** *Students know* that magnets have two poles (north and south) and that like poles repel each other while unlike poles attract each other.

An assortment of magnets of different sizes and shapes can be used to demonstrate this standard. Two or more donut-shaped magnets may be strung in different order on a pencil, and one magnet may be suspended in the air. Students can observe the bouncing effect that results from the repulsive force increasing as the magnets get closer. Refrigerator magnets are made of tiny strips of alternating north and south poles next to each other. They appear to just stick and never repel. However, by carefully sliding one refrigerator magnet over the other, one can perceive the alternate effects of repulsion and attraction.

> **1. g.** *Students know* electrical energy can be converted to heat, light, and motion.

Electrical energy is partially converted to heat and light when it flows through wires because the wires resist this flow. For example, light and heat are produced when electricity flows through the filament of a light bulb. Electrical energy may also be converted to kinetic energy by the use of devices such as an electromagnet or an electric motor. The conversion of electrical energy to heat can be demonstrated by feeling the warmth generated in the coil of an electromagnet when the circuit is completed. Common everyday experiences include incandescent light bulbs that are too hot to touch because most of the electrical energy is used to make the filament hot. Students can also observe that when they light a bulb with a battery, converting electrical energy into light, the lit bulb becomes warm.

If the current flowing through a circuit is high, the wires need to be thick enough to carry the current without becoming too hot. Extension cords designed to be used for high-current appliances, such as space heaters and microwave ovens, are thicker than extension cords designed to be used for a desk lamp. In addition, household circuits are rated to carry only a limited amount of current. If too many electrical devices are plugged into a circuit at once, a problem that is often made worse by "outlet expanders," a house fire may result from overheating of the wires. A fire may also result if the insulation in a wire is frayed and two wires short-circuit. The purpose of fuses and circuit breakers in a house is to prevent overloading of circuits and overheating of wires. Demonstrations in the classroom should never be done by using household current; even a short-circuited battery can generate enough heat to burn students, so teachers must carefully choose and monitor laboratory demonstrations and experiments.

STANDARD SET 2. Life Sciences

Students in grade four have already learned about types of plants and animals that inhabit different biomes and will have a simple understanding of adaptation from studies in grades one and three. The standards in grade four help to refine students' understanding of ecological principles and prepare them to learn much more about the subject in grade six.

> **2. All organisms need energy and matter to live and grow. As a basis for understanding this concept:**
>
> **a.** *Students know* plants are the primary source of matter and energy entering most food chains.

A food chain is a representation of the orderly flow of matter and energy from organism to organism by consumption. Plants harness energy from the sun, herbivores eat plants, and carnivores eat herbivores. Solar energy therefore sustains herbivores and, indirectly, the carnivores that eat them; this is the important principle to be taught.

> **2. b.** *Students know* producers and consumers (herbivores, carnivores, omnivores, and decomposers) are related in food chains and food webs and may compete with each other for resources in an ecosystem.

Students may recall from previous grade levels that animals eat plants or other animals. This standard extends the subject to a greater depth. Food chains and food webs represent the relationships between organisms (i.e., which organisms are consumed by which other organisms). Generally, food chains and food webs must originate with a primary producer, such as a plant that is producing biomass. Herbivores and omnivores eat the plants; carnivores (secondary consumers) in turn eat the herbivores and omnivores. Decomposers consume plant and animal waste, a step that returns nutrients to the soil and begins the process again. Decomposers, such as fungi and bacteria, should be included at each level of the food web as they consume the remains and wastes of plants and animals.

> **2. c.** *Students know* decomposers, including many fungi, insects, and microorganisms, recycle matter from dead plants and animals.

Plant and animal wastes, including their dead remains, provide food for decomposer organisms such as bacteria, insects, fungi, and earthworms. Decomposers are adept at breaking down and consuming waste materials and therefore complete the food chain, returning nutrients to the soil so that plants may thrive as producers. Bacteria and fungi also pass energy to other parts of a food web. Those microorganisms are themselves consumed by slightly larger organisms, such as worms and small insects, and those small consumers are food for larger animals, such as birds. Microorganisms and their biological ability to decompose matter may be observed in video or film productions using time-lapse photography. Molds grown on bread and fruit may be studied with the use of magnifying lenses; however, it is dangerous for a class to collect wild fungi, culture bacteria, or molds derived from soils or rotting meats.

STANDARD SET 3. Life Sciences

Students have learned in previous grades about the interactions of organisms in an ecosystem; this standard set develops the subject still further. The living and nonliving components are clearly distinguished, and the significant effects of invisible microorganisms are also discussed.

> **3. Living organisms depend on one another and on their environment for survival. As a basis for understanding this concept:**
>
> **a.** *Students know* ecosystems can be characterized by their living and nonliving components.

Each ecosystem is characterized by a set of living (biotic) and nonliving (abiotic) components that distinguish it from other ecosystems. For example, tropical rain forests, coral reefs, and deserts all have distinctly different biotic and abiotic components. This standard challenges students to be systematic in describing the components of an ecosystem and in identifying the characteristics of life.

3. b. *Students know* that in any particular environment, some kinds of plants and animals survive well, some survive less well, and some cannot survive at all.

This standard is partly an extension of the study of adaptive characteristics of plants and animals that students may have encountered in grade three. All living organisms have biological requirements for growth and survival and can live only in environments to which they are well adapted. If an environment changes in a way that is harmful to an organism, the organism may not be able to survive. Adaptation is a genetic process that takes many generations to be perceived, so a single individual cannot "adapt" to a change. For example, the thick, blubbery skin of whales is an evolutionary adaptation to cold water. This adaptation is different from the types of changes that help a single individual survive, such as a change in seasonal diet or coloration, which are properly called *accommodations.*

3. c. *Students know* many plants depend on animals for pollination and seed dispersal, and animals depend on plants for food and shelter.

The idea of plants and animals being mutually dependent was a topic of discussion in grade one. The concept can now be discussed at a much deeper level because students will have an emerging grasp of ecology and natural history. Many plants depend on bees, birds, and bats to pollinate their flowers. The resulting seeds may be scattered away from the parent plant by becoming entangled in the fur of animals. Other seedpods are moved and stored by animals in seed caches; some are consumed and deposited (still fertile) in animal wastes. The fruits of some plants are attractive food sources for animals. Plants often provide shelter for animals, hiding them from predators.

3. d. *Students know* that most microorganisms do not cause disease and that many are beneficial.

Microorganisms play a vital role in the environment. This standard helps students to look beyond the common misconceptions that bacteria are responsible only for diseases and that microorganisms are responsible only for decomposition. Some bacteria and single-celled organisms called *protists* are photosynthetic, and their contribution as primary producers of biomass in the ocean far exceeds that of the "visible" plants. Food chains and food webs may be based on bacteria and protists; therefore, a microscope will help students to observe microorganisms.

Growing cultures in the classroom provides students with opportunities to study bacteria and protists. A hay infusion is relatively safe to grow in a classroom. Within a few days students will be able to see numerous types of microorganisms through a microscope. Teachers and students should not culture soils and meat broths as some microorganisms can cause serious illness.

STANDARD SET 4. Earth Sciences (Rocks and Minerals)

Earth sciences standards in grade four are divided into three areas of study: rocks, minerals, and the processes of erosion. The topics extend what students have already learned in grade two and prepare them for a deeper level of understanding in grade six.

4. The properties of rocks and minerals reflect the processes that formed them. As a basis for understanding this concept:

a. *Students know* how to differentiate among igneous, sedimentary, and metamorphic rocks by referring to their properties and methods of formation (the rock cycle).

Rocks are usually made from combinations of different minerals and are identified from their composition and texture. Molten magma and lava cool and solidify to form igneous rocks. Metamorphic rocks form when a parent rock of any type is subjected to significant increases in pressure and temperature, short of melting. Sedimentary rock forms when rock is weathered, transported by agents of erosion, deposited as sediment, and then converted back into solid rock—a process called *lithification.* For classroom discussions it is best to begin with minerals and then progress to rocks. (See the next standard for a discussion of teaching about minerals.)

Students learn to sort rock specimens into groups of igneous, sedimentary, and metamorphic rocks. Students should learn to relate descriptions of rock mineral content and properties to the three rock groups. Rocks that are hard but show no layering are likely to be igneous rocks. Often they have interlocking crystalline textures. Rocks that are soft, particularly those with layers, are likely to be sedimentary rocks. They often have "fragmental" textures; they look like broken grains of older rocks cemented back together. Hard rocks that have their minerals lined up or arranged in uneven layers are likely to be metamorphic rocks. This description briefly depicts some of the most common rocks; however, there are many exceptions. Field guides to rocks and minerals may be checked out from the school library-media center and would be helpful to have for reference in the classroom.

4. b. *Students know* how to identify common rock-forming minerals (including quartz, calcite, feldspar, mica, and hornblende) and ore minerals by using a table of diagnostic properties.

Geologists describe and identify minerals according to a set of properties, such as hardness, cleavage, color, and streak. Hardness is determined by the Mohs hardness scale, which refers to materials' relative ability to scratch other materials or be scratched by them. Most earth sciences books contain tables of diagnostic mineral properties that can be used to assist students with sorting or classifying minerals.

The identification process requires matching the observed properties of a sample with those noted on a diagnostic table of properties. This standard focuses on only a few of the most common rock-forming minerals (e.g., quartz, calcite, feldspar, mica, hornblende) as well as some important ores, such as galena (lead) and hematite (iron). The colorful ores of copper may also be added to this list. Other resources, such as field guidebooks, computer programs, approved and preselected Internet sources, and resources from the school library, may help students identify mineral samples.

STANDARD SET 5. Earth Sciences (Waves, Wind, Water, and Ice)

The processes of weathering and erosion continually form the sediments that form new rocks as a part of the constant recycling of Earth's crust. Some changes on Earth's surface take place so slowly that they are hard for students to observe; others occur so rapidly that they may be frightening. Erosion and transportation are processes in which material is transported over short or long distances and may take place at different rates. Movement along faults may be slow or fast. Earth's surface may be built up slowly or erupt suddenly. Students tend to overemphasize the effectiveness of rapid processes because they are easy to identify, but the slow processes may ultimately have the greatest effect on the shape of Earth's surface.

> **5. Waves, wind, water, and ice shape and reshape Earth's land surface. As a basis for understanding this concept:**
>
> **a.** *Students know* some changes in the earth are due to slow processes, such as erosion, and some changes are due to rapid processes, such as landslides, volcanic eruptions, and earthquakes.

Erosion may occur so slowly that careful measurements are necessary to establish that a change is taking place; however, landslides may take place very rapidly. Volcanoes can build with explosive speed and then be quiet for long periods. Breaks in Earth's crust, called *faults*, experience slow movement, called *creep*, and rapid movements that cause earthquakes.

> **5. b.** *Students know* natural processes, including freezing and thawing and the growth of roots, cause rocks to break down into smaller pieces.

Chemical weathering occurs when atmospheric components (e.g., oxygen, carbon dioxide, and water) interact with Earth's surface materials and cause them to break apart or dissolve. Purely physical processes, such as alternate freezing and thawing of water, exfoliation, or abrasion, may also contribute to the weathering process. Plants may promote weathering as their roots expand in cracks to break rocks. Weathering results in the formation of soil or sediment.

To demonstrate the effects of freezing and thawing, teachers may use plastic bottles. Students can fill a soft plastic bottle with cold water. They make certain that all the air is removed from the bottle before tightly capping it and placing it in a freezer. The expansion of water as it freezes will deform the bottle and possibly even split it.

> **5. c.** *Students know* moving water erodes landforms, reshaping the land by taking it away from some places and depositing it as pebbles, sand, silt, and mud in other places (weathering, transport, and deposition).

Weathering produces pebbles, sand, silt, and mud. Erosion and transportation move the products of weathering from one place to another. As erosion transports the broken and dissolved products of weathering, it alters the shape of landforms. The most important agent of transportation is water. Water flowing in streams is energetic enough to pick up and carry silt, sand, pebbles, mud, or at flood stage even boulders. Flowing water reshapes the land by removing material from one place and depositing it in another.

STANDARD SET 6. Investigation and Experimentation

Students in grade four improve their ability to recognize the difference between evidence and opinion. They learn the difference between observation and the inference of some underlying cause or unseen action. Teachers will need carefully designed investigations and experiments that result in predictable student errors to teach the students the difference between observation and inference. Another important milestone is that students will learn to formulate cause-and-effect relationships and to connect predictions and results.

> **6. Scientific progress is made by asking meaningful questions and conducting careful investigations. As a basis for understanding this concept and addressing the content in the other three strands, students should develop their own questions and perform investigations. Students will:**
>
> **a.** Differentiate observation from inference (interpretation) and know scientists' explanations come partly from what they observe and partly from how they interpret their observations.
>
> **b.** Measure and estimate the weight, length, or volume of objects.
>
> **c.** Formulate and justify predictions based on cause-and-effect relationships.
>
> **d.** Conduct multiple trials to test a prediction and draw conclusions about the relationships between predictions and results.
>
> **e.** Construct and interpret graphs from measurements.
>
> **f.** Follow a set of written instructions for a scientific investigation.

Chapter 3
The Science Content Standards for Kindergarten Through Grade Five

Grade Five — Science Content Standards

Students in grade five will learn about chemical reactions and discover the special (and shared) properties of metallic elements. They will clearly distinguish between molecules and atoms and chemical compounds and mixtures and learn about the organization of atoms on the periodic table of the elements. They can then be shown how particular chemical reactions (e.g., photosynthesis and respiration) drive the physiological processes of living cells. They will add to what they have learned in previous grade levels about the external characteristics and adaptations of plants and animals and learn about some of the fundamental principles of physiology. They will learn about blood circulation and respiration in humans; digestion of food and collection and excretion of wastes in animals; the movement of water and minerals from the roots of plants to the leaves; and the transport of sugar generated during photosynthesis from the leaves to the other parts of the plant.

Students in grade five also study the hydrologic cycle (water cycle), the process by which water moves between the land and the oceans. They will learn how the hydrologic cycle influences the distribution of weather-related precipitation and, as a consequence, the types and rates of erosion. They will also study the solar system and learn that it contains asteroids and comets in addition to the Sun, nine planets,* and moons. They will learn the composition of the Sun and the relationship between gravity and planetary orbits.

The *Science Content Standards* and *English–Language Arts Content Standards* are complementary so that the writing strategies will lay a foundation for good writing on science reports and informative oral science presentations.[3] The *Science Content Standards* and the *Mathematics Content Standards* also reinforce each other as students analyze, strategize, and solve problems, finding solutions to apply to new circumstances. Students in grade five will also develop testable questions and learn to plan their own investigations, selecting appropriate tools to make quantitative observations.

* Under resolutions passed by the International Astronomical Union on August 24, 2006, there are eight planets. Pluto no longer meets the definition of a "planet" but is now classified under a new distinct class of objects called "dwarf planets."

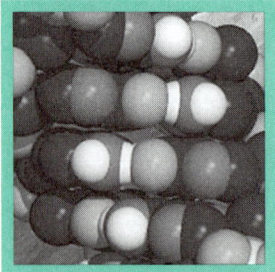

STANDARD SET 1. Physical Sciences

Students will have some familiarity with the idea of atoms and elements from science studies in grade three. In grade five the introduction to chemical reactions and the concept that atoms combine to form molecules require students to clearly distinguish between molecules and atoms and chemical compounds and mixtures. They will be introduced to the idea that the organization of atoms on the periodic table of the elements is related to similarities and trends in the chemical properties of the elements.

> **1. Elements and their combinations account for all the varied types of matter in the world. As a basis for understanding this concept:**
>
> **a.** *Students know* that during chemical reactions the atoms in the reactants rearrange to form products with different properties.

The properties of a chemical compound are controlled by the way atoms of different elements combine to make the compound. During a chemical reaction between two compounds, none of the original atoms are lost, but the atoms rearrange themselves into new combinations, resulting in the formation of products with properties that differ from those of the reacting compounds. Simple and safe chemistry experiments are described in fifth-grade science texts, and students can identify reactants and products when observing chemical reactions.

> **1. b.** *Students know* all matter is made of atoms, which may combine to form molecules.

The fact that atoms can combine to form molecules is new information, and students should be given the opportunity to practice the correct use of those terms. The number of different types of atoms is relatively small in comparison with the large number of different types of molecules that may be formed. Simple molecules (such as nitrogen, oxygen, water, carbon dioxide, methane, and propane) can be easily represented by molecular models, and this depiction can enhance students' understanding of the symbolic representations in text. The idea of combinations of atoms sets the stage for learning about chemical bonds in high school.

> **1. c.** *Students know* metals have properties in common, such as high electrical and thermal conductivity. Some metals, such as aluminum (Al), iron (Fe), nickel (Ni), copper (Cu), silver (Ag), and gold (Au), are pure elements; others, such as steel and brass, are composed of a combination of elemental metals.

Elements are grouped together on the periodic table of the elements according to their chemical properties, which in turn are based on the atomic structure of those elements. All pure, elemental metals share some properties in common, such as high electrical and thermal conductivity. Those same properties persist when elemental metals are combined to form alloys (e.g., copper and zinc to make brass).

Students may be familiar with many metallic elements (e.g., gold, silver, copper, zinc, aluminum, lead, mercury, chromium) and common metal alloys (e.g., brass, steel, bronze, pewter). It would be helpful for teachers to obtain samples of some of these metals and alloys for their students to study. (*Caution*: Some heavy metals [such as lead, mercury, and chromium, or their salts] may be hazardous.) In general, metals are shiny, reflecting most of the light that strikes them. They are malleable and ductile (that is, they will bend under pressure and are not brittle). They have a broad range of melting temperatures (e.g., mercury is a liquid at room temperature, gallium will melt in one's hand, and tungsten has a melting temperature around 3,400 degrees Celsius). The thermal and electrical conductivity of all metals is high compared with nonmetallic substances, such as plastics and ceramics, rocks, and solid salts. Given the appropriate tools, students can develop tests for metals and nonmetals to determine whether they conduct electricity and heat.

> **1. d.** *Students know* that each element is made of one kind of atom and that the elements are organized in the periodic table by their chemical properties.

All matter is made of atoms. The word *element* refers to those substances that repeated experiments have shown cannot be reduced to still more "elementary" substances. The explanation for this fact is that elements are made of many identical atoms. Water was considered an element at one time. However, it is possible to electrolyze water and produce hydrogen and oxygen gas, both elements. The properties of elements are determined entirely by their atoms. Therefore, elements are said to be made of one kind of atom that accounts for the element's unique properties. The history of the discovery and name of any one of the elements provides insight into the nature of science and scientific progress. The single most important property of an element is its atomic number. The number may be found on the periodic table along with the symbol and name of the element. Students should know that atomic numbers increase as they read from left to right and move line by line down the periodic table.

In grade eight students will be taught that the physical and chemical properties of an element are based on the internal structure of its atoms. The periodic table was originally constructed on the basis of increasing atomic weights of the elements. Those elements were organized in the pattern of a table, much like a monthly calendar, so that elements with similar chemical properties (e.g., metals, halogens, and noble gases) are grouped together in columns. The table gets its name because of the repeating, or periodic, sequences of chemical properties. Students should examine the periodic table of the elements and be able to locate elements by name. They should be able to find common metallic elements on the table and learn to refer to the table as they study and experiment with substances whose names are composites of the elements, such as sodium chloride and carbon dioxide.

> **1. e.** *Students know* scientists have developed instruments that can create discrete images of atoms and molecules that show that the atoms and molecules often occur in well-ordered arrays.

The technique of electron microscopy has opened the door to a new generation of analytical tools that can be used to produce images of individual atoms in a crystalline array. Those images show atoms as "fuzzy balls" aligned in orderly and repeating patterns. From those images it is possible to infer that atoms are discrete objects of finite size and nearly spherical shape. Students may see images in textbooks and on the Internet that were obtained by using atomic-resolution instruments, such as electron microscopes and scanning tunneling microscopes. Those images confirm, as hypothesized from years of indirect experimental evidence, that atoms in metals and crystals are arranged in orderly array. The images also show the presence of microfractures in which the order is interrupted, a condition that can affect the strength of the material.

> **1. f.** *Students know* differences in chemical and physical properties of substances are used to separate mixtures and identify compounds.

Students should know the difference between mixtures and compounds. In compounds atomic constituents are separated by chemical rather than by physical means. In addition, every compound has a unique set of chemical and physical properties that can be used to identify it. Compounds and classes of compounds may be identified by chemical reactions with other compounds. An example is the iodine starch reaction. Other chemical reactions in solution may be explored to identify compounds based on changes in acidity, formation of precipitates, and changes in color. In mixtures the atomic constituents are separated by their physical properties. Simple and safe activities may be found in science texts for students in grade five. For example, iron filings can be separated from nonmetallic materials by use of a magnet, and a piece of filter paper can be used to separate suspended particles in a solution.

> **1. g.** *Students know* properties of solid, liquid, and gaseous substances, such as sugar ($C_6H_{12}O_6$), water (H_2O), helium (He), oxygen (O_2), nitrogen (N_2), and carbon dioxide (CO_2).

This standard builds on the previous one by challenging students to describe and identify a few common elements and compounds on the basis of observed chemical properties. Students can also study the three common physical states of matter for each of these compounds or elements as well as learn about and compare such properties as solubility in water, boiling and freezing points, sublimation, and reactivity.

> **1. h.** *Students know* living organisms and most materials are composed of just a few elements.

By weight 98.59 percent of Earth's entire crust consists of eight elements: oxygen, silicon, aluminum, iron, calcium, sodium, potassium, and magnesium. Nearly 3,500 known minerals are in Earth's crust. This fact shows that the complexity of the crust is also the result of a small number of elements in a large variety of combinations. Similarly, living organisms are mostly composed of the elements carbon, oxygen, hydrogen, nitrogen, sulfur, and phosphorus. The number of types of atoms used as "building blocks" is relatively small. The way in which the atoms are organized into molecules provides variety.

1. i. *Students know* the common properties of salts, such as sodium chloride (NaCl).

Elements and compounds may be described and identified on the basis of observed chemical and physical properties. Salts are compounds typically made from a metal and a nonmetal. Many salts are hard and brittle and have high melting temperatures. Most salts are soluble in water. When dissolved, they become conductors of electricity.

Salts are made when strong acids react with strong bases. For example, in the reaction of hydrochloric acid (HCl) with sodium hydroxide (NaOH), hydrogen (H) combines with hydroxide (OH) to form water while sodium (Na) and chlorine (Cl) ions remain in a solution that, if evaporated, would leave the salt sodium chloride (NaCl). Although the use of strong acids and bases in elementary classrooms would present a significant safety risk, science materials adopted for instruction in grade five describe simple experiments that can be safely conducted.

There are many different types of salts, but the general use of the term *salt* refers to sodium chloride, the most common and widely used. In science many salts are (but are not limited to) substances formed by elements in the groups under sodium and magnesium in combination with elements under fluorine. Some salts are poisonous, and students need to be cautioned not to ingest any substances used or produced in an experiment.

STANDARD SET 2. Life Sciences

In grade one students were presented with a simple example of the relationship between structure and function; namely, that the shapes of teeth are related to the types of materials animals eat. They subsequently learned to identify this phenomenon as an *adaptation.* Much of the discussion to this point has focused on external characteristics, but plants and animals have internal structures as well that perform vital functions. This subject, which is commonly called physiology, is developed still further in grade seven and in high school.

2. Plants and animals have structures for respiration, digestion, waste disposal, and transport of materials. As a basis for understanding this concept:

   a. *Students know* many multicellular organisms have specialized structures to support the transport of materials.

Multicellular organisms usually have cells deep within them that need to receive a supply of food and oxygen and, in the case of animals, to have cellular wastes removed. In higher-order animals blood circulation is responsible for transporting glucose sugar to each cell, providing oxygen, and removing cellular wastes and carbon dioxide. To demonstrate the transport of water in a plant, the teacher may cut the bottom end of a stalk of celery and place it in water containing food coloring. After the colored water is taken up into the plant, students can make cross-sections of the celery and observe them under a microscope. Observing the cross-sections is helpful to students in understanding Standard 2.e.

> **2. b.** *Students know* how blood circulates through the heart chambers, lungs, and body and how carbon dioxide (CO_2) and oxygen (O_2) are exchanged in the lungs and tissues.

Structures of the cardiovascular and circulatory systems, including the heart and lungs, promote the circulation of blood and exchange of gas. The left side of the heart is responsible for pumping blood through arteries to all the tissues of the body and delivering oxygen. Oxygen-poor blood returns to the heart through veins; the right side of the heart is responsible for pumping this blood to the lungs, where the blood eliminates its carbon dioxide and receives a fresh supply of oxygen. Exhaling expels the carbon dioxide that was transported to the lungs by the blood; inhaling allows the intake of oxygen, which is picked up by the blood.

> **2. c.** *Students know* the sequential steps of digestion and the roles of teeth and the mouth, esophagus, stomach, small intestine, large intestine, and colon in the function of the digestive system.

Digestion starts in the mouth, where chewing breaks down food into smaller pieces that can be easily swallowed and digested. Saliva contains compounds that are also important in breaking down food. The esophagus is a tube that moves food from the mouth to the stomach after swallowing. In the stomach the food is mixed with stomach acids that help to break down the food into parts that can be absorbed. Once food reaches the small intestine, it is neutralized and processed into molecules that can be absorbed into the blood supply. The large intestine recovers water from food, and the colon collects fecal waste (indigestible parts of food) and stores it prior to elimination from the body.

> **2. d.** *Students know* the role of the kidney in removing cellular waste from blood and converting it into urine, which is stored in the bladder.

Cells in living organisms produce waste products that they cannot recycle into other compounds. The focus of this standard is on the systems that remove waste from the cells to prevent it from accumulating and eventually poisoning the organism. Cellular waste products (in the form of molecules) are separated from the bloodstream by the kidneys, stored in the bladder as urine, and removed from the body by urination. In plants many such waste products are stored in a large central vacuole in each plant cell—a kind of garbage dump that is gradually filled as the cell ages.

> **2. e.** *Students know* how sugar, water, and minerals are transported in a vascular plant.

The *xylem* of plants is a woody tissue responsible for water and mineral transport from roots to leaves. Water moving up the plant stem replaces water that has evaporated from the leaves. Plants also transport sugar from the leaves to the roots through a living structure of tubes called the *phloem*.

> **2. f.** *Students know* plants use carbon dioxide (CO_2) and energy from sunlight to build molecules of sugar and release oxygen.

Photosynthesis is the name of the process by which plants capture the energy of the sun and use it to initiate a chemical reaction between carbon dioxide and water that results in the production of sugar molecules and the release of oxygen molecules. The chemical process is as follows:

energy + carbon dioxide + water react to form sugar + oxygen

The process is expressed in the following equation:

$$\text{energy} + 6\,CO_2 + 6\,H_2O \rightarrow C_6H_{12}O_6 + 6\,O_2$$

The sugar made during photosynthesis is just an initial compound the plant produces. All the other organic molecules are made by modification of this simple compound. For example, a significant portion of the mass of a log from a tree was once carbon dioxide gas in the air, captured by the leaves of a tree, and fixed into larger organic molecules as shown by the equation noted above. The sugar transport processes in the tree are also important in moving the products of photosynthesis down to the stem, where they could then become a part of the tree.

> **2. g.** *Students know* plant and animal cells break down sugar to obtain energy, a process resulting in carbon dioxide (CO_2) and water (respiration).

Cellular respiration is a process of producing energy by the chemical breakdown of carbohydrate (sugar) molecules—a process that is the reverse of photosynthesis. The chemical process is as follows:

sugar + oxygen react to form carbon dioxide + water

The process is expressed in the following equation:

$$C_6H_{12}O_6 + 6\,O_2 \rightarrow 6\,CO_2 + 6\,H_2O$$

Both plants and animals break down sugar to release its energy in a form they can use. This process is called *cellular respiration*. Carbon dioxide and water are reaction by-products. In animals the carbon dioxide is released into the blood, where it can be transported to the lungs. In the lungs carbon dioxide and oxygen are exchanged (which is the other use of the term *respiration*) during the act of breathing. It should be noted that cellular respiration is not the same as breathing.

STANDARD SET 3. Earth Sciences (Earth's Water)

The hydrologic cycle (water cycle) is the process by which water moves between the land and the oceans. Students in grade five learn that cooling in the atmosphere returns water vapor to a liquid or solid state as rain, hail, sleet, or snow. They are also introduced to factors that control clouds, precipitation, and other weather phenomena.

> **3. Water on Earth moves between the oceans and land through the processes of evaporation and condensation. As a basis for understanding this concept:**
>
> **a.** *Students know* most of Earth's water is present as salt water in the oceans, which cover most of Earth's surface.

Because water covers three-fourths of Earth's surface, this planet is sometimes referred to as the blue planet. Fresh water falls as rain on land and oceans alike. When it falls on land, the water dissolves salts and other mineral matter and carries them to the oceans. When water evaporates from the surface of the ocean, the salts remain behind and accumulate. For this reason the oceans have become salty. Students should know that the amount of fresh water on land is small compared with the amount in the oceans. Using science texts aligned with the *Science Content Standards* or a variety of library and other resources, students should be able to trace diagrams of the water cycle and understand what they represent.

> **3. b.** *Students know* when liquid water evaporates, it turns into water vapor in the air and can reappear as a liquid when cooled or as a solid if cooled below the freezing point of water.

Liquid water evaporates and becomes invisible vapor when warmed by the sun. Water vapor mixes with the air as it moves through the atmosphere. When the air is cooled, a fraction of the water vapor changes back to liquid water in the form of clouds or rain. If the air temperature becomes low enough, the water will crystallize into a solid state as snow, sleet, or hail. Alternating periods of evaporation and precipitation drive the hydrologic cycle. For a laboratory demonstration a teacher may boil water to produce water vapor and direct the vapor onto the cold outside surface of a beaker filled with ice water. The precipitated water vapor will fog the outside of the beaker with tiny drops of liquid water.

> **3. c.** *Students know* water vapor in the air moves from one place to another and can form fog or clouds, which are tiny droplets of water or ice, and can fall to Earth as rain, hail, sleet, or snow.

Atmospheric circulation moves water vapor, clouds, and fog from one place to another. The tiny droplets or crystals of water that form fog and clouds are so small that they remain suspended in the air. Further cooling of the air can cause these droplets or crystals to grow sufficiently until they fall to the earth as rain, hail, sleet, or snow. By learning basic meteorology from texts and monitoring and plotting local weather data reported by the news media, students can explore the relationship between the amount of water vapor in the air (humidity), air temperature, and the likelihood of rainfall or snowfall.

> **3. d.** *Students know* that the amount of fresh water located in rivers, lakes, underground sources, and glaciers is limited and that its availability can be extended by recycling and decreasing the use of water.

Students learn that water quality is affected by various uses and that there are local, state, federal, and global efforts to manage water resources. In California, water resources depend on the use of annual rainwater (and snowpack water) collected in watershed districts, pumping of groundwater, import of water from rivers, and reclamation of water that has been used. Water quality in streams is affected by the disturbance or development of land in a watershed area, runoff of water from farms and city streets, and projects that control the flow of rivers in a flood basin.

> **3. e.** *Students know* the origin of the water used by their local communities.

Students learn the origins of the local water supply through a study of the watershed, creeks, rivers, aqueducts, dams, and reservoirs that serve as its source. Students should know whether their community's balance between water supply and demand varies seasonally and whether conservation and reclamation techniques are practiced. If water is imported, students should be able to trace it back to its source or sources.

STANDARD SET 4. Earth Sciences (Weather)

Students in grade five learn about the causes of large-scale and small-scale movements in the atmosphere. They apply knowledge of the hydrologic cycle to understanding weather and weather patterns.

> **4.** Energy from the Sun heats Earth unevenly, causing air movements that result in changing weather patterns. As a basis for understanding this concept:
>
> a. *Students know* uneven heating of Earth causes air movements (convection currents).

The atmosphere and surface of Earth are heated unevenly, giving rise to both local and global temperature differences. For example, the direct heat absorbed by the surface of the ocean, land, and air may result in different temperatures. Furthermore, the amount of heat varies with latitude, primarily because of the height of the Sun in the sky. The lower the Sun's elevation, the less direct is its radiation and the less radiation that falls on each square meter of Earth's surface area. This event is a result of geometry and depends on the angle at which the Sun's rays intersect Earth's surface at a locality. When the incoming rays of the Sun intersect Earth's surface at a more oblique angle, the solar flux is spread out over a wider area. Polar regions are cold because the Sun is low in the sky and its rays fall at very large angles. Closer to the equator, the Sun's rays fall more directly and the climate is hot. The uneven heating results in local and global temperature differences that create convection currents in the oceans and atmosphere. Students in grade five should know that warm air rises and cold air falls toward Earth's surface, setting up convection currents in the air that are called *winds.*

Convection is an important mechanism in moving heat around in Earth's mantle, in the oceans, and in the atmosphere. The process of hot air rising and cold air sinking occurs at Earth's surface on many different scales, causing local winds and great global air currents, such as the trade winds.

4. b. *Students know* the influence that the ocean has on the weather and the role that the water cycle plays in weather patterns.

Because Earth is a sphere, equatorial regions receive more concentrated sunlight than do polar regions. Temperatures are therefore higher at the equator than farther north or south, but the difference would be much more extreme without the influence of the oceans, which cover about 70 percent of Earth's surface. Large bodies of water can absorb (or release) a great deal of heat without changing temperature very much; their temperature stays relatively constant from day to night and from season to season. Oceanic circulation carries water warmed near the equator to the north and to the south. The great ocean currents help distribute heat from place to place by gradually releasing it into Earth's atmosphere. Warm surface currents (such as the Gulf Stream) make high-latitude countries (such as Scotland) more habitable than they would otherwise be. Moreover, a great amount of equatorial heat is absorbed by water during evaporation. Global atmospheric currents (winds) carry the water vapor to cooler regions, and heat is released to the atmosphere as the vapor condenses, forming precipitation. Thus heat as well as water is transported, providing an important mechanism for evening out temperatures on Earth.

Air in contact with large bodies of water is *tempered*—warmed in the winter and cooled in the summer. The amount and distribution of precipitation depend a great deal on the surface temperature of the water. When water temperatures do change, even a little, large changes in weather patterns may occur. A good example of this is the ENSO (El Niño/Southern Oscillation) cycle, which brings especially wet and dry seasons to many places around the world.

> **4. c.** *Students know* the causes and effects of different types of severe weather.

Many types of severe weather are in the world: hurricanes, tornadoes, thunderstorms, and monsoons. The source of energy for all weather is the Sun, which heats air and water unevenly. Warm air tends to be less dense than cold air, and air will always flow (blow) from areas of high pressure (denser air) toward areas of lower pressure, creating winds. With increasing temperature, more water can evaporate into the air. When this warm, moist air is suddenly cooled (as by contact with a cold air mass), condensation and precipitation may result. The contacts between air masses with different temperatures are called *fronts*. When a patch of warm, low-pressure air is surrounded by higher-pressure air (called a low-pressure "closure"), the warmer air will tend to rise and be replaced, through convection, by high-pressure air flowing in from all around. Because Earth rotates on its axis, all such winds are deflected (turned to the right in the Northern Hemisphere and to the left in the Southern); the net effect is a circular wind, which surrounds the low-pressure closure. The rising warm air in the center cools, its water condenses, and precipitation occurs. This phenomenon is known as a *cyclone* and is the cause of many big hurricanes and other storms.

> **4. d.** *Students know* how to use weather maps and data to predict local weather and know that weather forecasts depend on many variables.

Weather maps display data on air temperature, air pressure, and precipitation. If students know that air flows from regions of high pressure to regions of low pressure (and turns to the right in the Northern Hemisphere), they can look at a weather map and predict the direction of the wind. If they know, for example, that weather fronts tend to move from west to east in North America, they can predict tomorrow's weather in one place by checking on today's weather somewhere else. And if they see low-pressure closures (discussed above), they can predict stormy or fair weather from high-pressure closures. Very small changes in temperature and pressure, however, may significantly change all such patterns over a few days (the so-called chaos theory). Long-term weather forecasts tend to be unreliable for this reason.

> **4. e.** *Students know* that the Earth's atmosphere exerts a pressure that decreases with distance above Earth's surface and that at any point it exerts this pressure equally in all directions.

Atmospheric pressure is the weight of air (a force) pushing on a given square unit area (e.g., m^2 or cm^2). Air is invisible, hard to detect by the sense of touch, and difficult to weigh. Thinking of air as being able to exert pressure works against one's intuition; nonetheless, air has mass and anything with mass is pulled by gravity toward Earth's center. This principle means that atmospheric pressure is greatest near Earth's surface at sea level and diminishes with increasing height in the atmosphere.

This effect is used by airplane pilots to measure altitude reliably, with barometric pressure at sea level serving as a reference point. The principle also means the pressure exerted on the bottom of an object, such as a balloon, is slightly greater than the pressure on the top. The second part of this standard is a reminder that the direction of the "push" caused by the pressure is the same in all directions—up, down, or sideways. The same principle holds true for pressure in any fluid.

STANDARD SET 5. Earth Sciences (The Solar System)

Student knowledge of the solar system includes an understanding of and the ability to describe the relative motions of the planets. Students already know that Earth orbits the Sun and the Moon orbits Earth. Students in grade five learn the composition of the Sun and that the solar system includes small bodies, such as asteroids and comets, as well as the Sun, nine planets,* and their moons. They learn the basic relationship between gravity and the planetary orbits.

5. The solar system consists of planets and other bodies that orbit the Sun in predictable paths. As a basis for understanding this concept:

   a. *Students know* the Sun, an average star, is the central and largest body in the solar system and is composed primarily of hydrogen and helium.

The Sun is about one million times the volume of Earth. Its mass can be calculated from the shapes of the planetary orbits, which result from the gravitational attraction between the Sun and its planets. The fusion of hydrogen to helium produces most of the Sun's energy.

   5. b. *Students know* the solar system includes the planet Earth, the Moon, the Sun, eight other planets and their satellites, and smaller objects, such as asteroids and comets.

The solar system comprises nine planets,* in the following order from the Sun: Mercury, Venus, Earth, Mars, Jupiter, Saturn, Uranus, Neptune, and Pluto. Most of the planets have moons in orbit about them, but only Earth's moon is visible to the unaided eye. Asteroids and comets are small bodies, most of which are in irregular orbits about the Sun. Many science texts and Web sites provide information and photographs of objects in the solar system that are collected from NASA's planetary, comet, and asteroid missions and from the use of Earth and space telescopes.

   5. c. *Students know* the path of a planet around the Sun is due to the gravitational attraction between the Sun and the planet.

Planets move in elliptical but nearly circular orbits around the Sun just as the Moon moves in a nearly circular orbit around Earth. Each object in the solar system

---

* Under resolutions passed by the International Astronomical Union on August 24, 2006, there are eight planets. Pluto no longer meets the definition of a "planet" but is now classified under a new distinct class of objects called "dwarf planets."

would move in a straight line if it were not pulled or pushed by a force. Gravity causes a pull, or attraction, between the mass (matter) of each of the planets and the mass (matter) of the Sun. This pull is what continually deflects a planet's path toward the Sun and produces its orbit.

Students may wonder why the pull of gravity does not cause the planets to "fall" into the Sun or the Moon into Earth. One explanation is that the planets and Moon are in fact falling, but they are also moving very fast to the side. As the Moon is pulled toward Earth, it also moves forward creating the curved path of its orbit. Thus the Moon is constantly falling, but the downward and sideways motions are exactly balanced so that the Moon never gets closer to or farther away from Earth. In the same way the planets are maintained in orbits around the Sun. Understanding that gravity exists in outer space may be made more difficult by the images of astronauts floating "weightless" in their capsules. When these pictures are taken, the astronauts are in orbit around Earth and are essentially free-falling (just like the Moon).

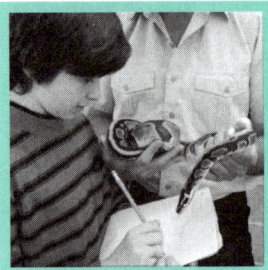

STANDARD SET 6. Investigation and Experimentation

Questions that are testable in science are founded on factual information and are based on observations. When students plan an experiment on the basis of their questions, they must decide what the variables are or what properties or sequence of events will change throughout the experiment. Students will observe and measure a change in one of the properties or event sequences in their experiment. The experiment is complete when the students draw conclusions and make inferences in a written or oral report or in both.

6. **Scientific progress is made by asking meaningful questions and conducting careful investigations. As a basis for understanding this concept and addressing the content in the other three strands, students should develop their own questions and perform investigations. Students will:**

   a. Classify objects (e.g., rocks, plants, leaves) in accordance with appropriate criteria.

   b. Develop a testable question.

   c. Plan and conduct a simple investigation based on a student-developed question and write instructions others can follow to carry out the procedure.

   d. Identify the dependent and controlled variables in an investigation.

   e. Identify a single independent variable in a scientific investigation and explain how this variable can be used to collect information to answer a question about the results of the experiment.

f. Select appropriate tools (e.g., thermometers, metersticks, balances, and graduated cylinders) and make quantitative observations.

g. Record data by using appropriate graphic representations (including charts, graphs, and labeled diagrams) and make inferences based on those data.

h. Draw conclusions from scientific evidence and indicate whether further information is needed to support a specific conclusion.

i. Write a report of an investigation that includes conducting tests, collecting data or examining evidence, and drawing conclusions.

Notes

1. *Science Safety Handbook for California Public Schools.* Sacramento: California Department of Education, 1999.
2. *Mathematics Content Standards for California Public Schools, Kindergarten Through Grade Twelve.* Sacramento: California Department of Education, 1999.
3. *Science Content Standards for California Public Schools, Kindergarten Through Grade Twelve.* Sacramento: California Department of Education, 2000; *English–Language Arts Content Standards for California Public Schools, Kindergarten Through Grade Twelve.* Sacramento: California Department of Education, 1998.

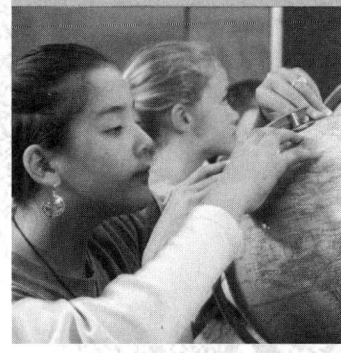

The Science Content Standards for Grades Six Through Eight

The Science Content Standards for Grades Six Through Eight

In each grade, kindergarten through grade five, the science content standards cover the areas of physical, life, and earth sciences in approximately equal measures. In each of the middle grades, however, the content standards emphasize an individual area. This organization permits students to probe each area in greater depth.

- In grade six the content standards focus on earth sciences. Students often become environmentally aware at this grade level, and this focus is meant to stimulate intellectual curiosity in that area.
- In grade seven the content standards focus on life sciences. Students at this grade level typically receive a semester of health education, and this focus is designed both to complement that instruction and to prepare students for the biology/life sciences course work that is often taken in the early high school years.
- In grade eight the content standards focus on physical sciences. This focus is designed to prepare students for the physics and chemistry course work that is often taken in the later high school years.

In all three of the middle grades, science instruction is intended to provide students with a solid foundation for the more formal treatment of concepts, principles, and theories called for at the high school level.

Not all students will enter middle school prepared for the rigorous science curriculum called for in the middle grades standards. Teachers should use "catch up" strategies to ensure that students are prepared for high school science. One of the key requirements is for students to have foundational reading and mathematics skills, as outlined in the State Board of Education's *Reading/Language Arts Framework for California Public Schools* and the *Mathematics Framework for California Public Schools*.[1] Those frameworks provide specific strategies for teachers to help students who are below grade level in reading and mathematics.

Students who are prepared to undertake the study of algebra (either as a separate course or as part of an integrated mathematics course) in grade eight, as called for in the *Mathematics Content Standards for California Public Schools*, will be on the pathway for success in high school science.[2] Those who are not as well prepared will struggle and may even fail in their science classes to the great frustration of their teachers and parents/guardians. For example, students who have not mastered arithmetic and algebra skills will find chemistry difficult, if not impossible. Science instruction should provide opportunities for students to use mathematics by solving problems. Teachers may use science to both reinforce mathematical abilities and deepen students' understanding of key mathematical concepts.

Safety is always the foremost consideration in the design of demonstrations, hands-on activities, laboratories, and science projects on site or away from school. Teachers must become familiar with the *Science Safety Handbook for California Public Schools*.[3] It contains specific and useful information relevant to classroom teachers of science. School administrators, teachers, parents/guardians, and students have a legal and moral obligation to promote safety in science education. Safety should be taught. Scientists and engineers in universities and industries are required to follow strict environmental health and safety regulations. Knowing and following safe practices in science are a part of understanding the nature of science and scientific enterprise.

Grade Six — Focus on Earth Sciences

The science curriculum in grade six emphasizes the study of earth sciences. Students at this age are increasing their awareness of the environment and are ready to learn more. The standards in grade six present many of the foundations of geology and geophysics, including plate tectonics and earth structure, topography, and energy. The material is linked to resource management and ecology, building on what students have learned in previous grades. Unless students take a high school earth science class, what they learn in grade six will be their foundation for earth science literacy.

STANDARD SET 1. Plate Tectonics and Earth's Structure

Plate tectonics is a unifying geologic theory that explains the formation of major features of Earth's surface and important geologic events. Although most scientists today consider Alfred Wegener to be the pioneer of the modern continental drift theory, he died with very little recognition for his accomplishment. Wegener asserted that evidence on Earth's surface indicated that the continents were once attached as an entire land mass. He theorized that this land mass broke up into pieces that subsequently drifted apart. Today, geologists know that plate tectonic processes are responsible for most of the major features of Earth's crust (including continental configuration, mountains, island arcs, and ocean floor topography) and are an important contributor to the recycling of material in the rock cycle. Driven by the flow of heat and material within Earth, these processes cause stresses in Earth's crust that are released through earthquakes and volcanic activity. Mountain building counters the constant destructive effects of weathering and erosion that eventually wear down Earth's surface features.

1. **Plate tectonics accounts for important features of Earth's surface and major geologic events. As a basis for understanding this concept:**

    a. *Students know* evidence of plate tectonics is derived from the fit of the continents; the location of earthquakes, volcanoes, and midocean ridges; and the distribution of fossils, rock types, and ancient climatic zones.

Evidence of past plate tectonic movement is recorded in Earth's crustal rocks, in the topography of the continents, and in the topography and age of the ocean floor. Continental edges reflect that they were once part of a single large supercontinent that Wegener named Pangaea. Upon the breakup of this supercontinent, the individual continents were moved to their present locations by the forces that drive plate tectonics. When the continental plates of today are returned to their supercontinent positions (through computer modeling), the fossil and sedimentary evidence of ancient life distributions and climate becomes coherent, providing strong support for the existence of Pangaea. As plates move in relation to one another, landforms and topographic features, such as volcanoes, mountains, valleys, ocean trenches, and midocean ridges, are generated along plate boundaries. Those regions are also frequently associated with geothermal and seismic activity. There is strong evidence that the divergence and convergence of the lithospheric plates did not begin and end with Pangaea but have been going on continually for most of the history of Earth.

Students should read and discuss expository texts that explain the process of continental drift and study maps that show the gradual movement of land masses over millions of years. Students may then model the process by cutting out continental shapes from a map of Earth and treating these continents as movable jigsaw puzzle pieces. Students read about the underlying evidence for continental drift and determine that the best-supported model of Pangaea shows a continuation of major geologic features and fossil trends across continental margins. The "broken" pieces of Pangaea can be gradually moved into their modern-day continental and oceanic locations. In doing this students should think carefully about the rate and time scale of the movement. This would be a good point in the curriculum to introduce the differing compositions of the denser ocean floor (basaltic) rock and less-dense continental (granitic) rocks. Students can also learn why most modern-day earthquakes and volcanoes occur at the "leading edges" of the moving continents.

> **1. b.** *Students know* Earth is composed of several layers: a cold, brittle lithosphere; a hot, convecting mantle; and a dense, metallic core.

Earth is not homogeneous solid rock but is composed of three distinct layers: a rocky, thin, fractured outer layer called the *crust;* a denser and thick middle layer called the *mantle;* and a dense, metallic center called the *core.* Geologists also use another classification scheme in which the outer, brittle layer of Earth is called the lithosphere (from the Greek *lithos* for rock). The lithosphere includes the crust and the outermost portion of the mantle and is the part that is broken into the tectonic plates. Students should know the properties of the crust, mantle, core, lithosphere, and plastic mantle region just beneath the lithosphere called the *asthenosphere.*

Temperature increases with increasing depth as a consequence of the heat released by the decay of trace quantities of radioactive atoms that are contained within Earth. Heating lowers the density of parts of the interior. Because an arrangement of high-density material over low-density material cannot be gravitationally stable, a vertical flow called *convection* develops. This convection can be sustained as long as the interior continues to be heated, causing a continuous cycling within Earth's interior. Scientists gather evidence for the details of Earth's layered structure from the analysis of seismic P and S waves as they pass through the planet. The content of this standard can be learned efficiently by the study of a cross-sectional model or diagram of Earth showing locations (to scale) of the crust, mantle, and core with each subdivision labeled according to temperature, density, composition, and physical state.

> **1. c.** *Students know* lithospheric plates the size of continents and oceans move at rates of centimeters per year in response to movements in the mantle.

Convective flow in the mantle moves at rates measured in centimeters per year, about as fast as fingernails grow. Mantle motion is transferred to the lithosphere at its boundary with the asthenosphere. As a result of this coupling, the lithospheric plates are carried passively along, riding as "passengers" at the same slow rate, in much the same way that ice floats along on slow-moving water. These lithospheric plates may be *oceanic* (i.e., they consist of rocks of basaltic composition) or *continental* (i.e., they consist of a more varied suite of rocks, mostly of granitic composition, covered in many places with a thin veneer of sedimentary rocks). Convective flow is based on the "rising" and "sinking" of materials with different relative densities. Just as a hot-air balloon rises through lower temperature and therefore denser air, hot convecting mantle can rise through lower temperature and therefore denser rock, albeit very slowly. When the material cools and increases in density, it may sink just like the hot-air balloon once the air inside has cooled. Oceanic lithosphere cools after it forms at Earth's surface and can eventually become dense enough that it will sink into Earth's interior, sliding under an adjacent plate that is less dense.

> **1. d.** *Students know* that earthquakes are sudden motions along breaks in the crust called faults and that volcanoes and fissures are locations where magma reaches the surface.

The hot, moving mantle is responsible for many geologic events, including most seismic and volcanic activity. As a result of the relative motion of the lithospheric plates, the boundaries of the plates are subjected to stresses. When the rocks are strained so much that they can no longer stretch or flow, they may rupture. This rupture is manifested as sudden movement along a broken surface and is called a *fault*. The energy released is spread as complex waves (called *earthquakes*) that travel through and around Earth. Volcanic phenomena, including explosive eruptions and lava flows, may also result from interactions at the boundaries between plates. Molten gas-charged magma generated in the crust or mantle rises buoyantly and exerts an upward force on Earth's surface. If these rocks and gases punch through the surface, they result in a variety of volcanic phenomena.

Using California-adopted texts, software, and other instructional materials aligned with the *Science Content Standards*, students can study models of the inner structures of volcanoes, the dynamics of the central crater, and the processes of erupting and flowing lava. Students can study the various types of volcanoes and how they form. They can also learn about different types of lava flows and the three major types of volcanic landforms (cone, shield, and composite).

> **1. e.** *Students know* major geologic events, such as earthquakes, volcanic eruptions, and mountain building, result from plate motions.

Most (but not all) earthquakes and volcanic eruptions occur along plate boundaries where the plates are moving relative to one another. The movement is never smooth; it may produce fractures or faults and may also generate heat. The sudden shift of one plate on another plate along faults causes earthquakes. Volcanic eruptions may occur along faults in which one plate slides under another and sinks deep enough to melt part of the descending material. This process of one plate sliding under another is called *subduction*. Great mountain-building episodes occur when two continental plates collide. The collision (although slow) is enormously powerful because of the mass of the continents. Over long periods of time, this process may crumple and push up the margins of the colliding continents.

Students may use a large map of the world or of the Pacific Ocean (including the entire Pacific Ocean Rim) to plot the locations of major earthquakes and volcanic eruptions during the past ten to 100 years. The locations of those tectonic events may be found on the Internet or in various library resources. Different symbols may be used to represent different depths or magnitudes of events. In studying such a map, students should note that tectonic events form a "ring" that outlines the Pacific Plate and that there is a Hawaiian "hot spot." Landforms associated with the plate boundaries include mountain belts, deep ocean trenches, and volcanic island arcs.

> **1. f.** *Students know* how to explain major features of California geology (including mountains, faults, volcanoes) in terms of plate tectonics.

Most of California resides on the North American lithospheric (continental) plate, one of the several major plates, and many smaller plates that together form the lithosphere of Earth. A small part of California, west of the San Andreas Fault, lies on the adjacent Pacific (oceanic) Plate. Geologic interactions between these two plates over time have created the complex pattern of mountain belts and intervening large valleys that make up the current California landscape. Large parts of the central and southern parts of California were once covered by a shallow sea. Interactions with the Pacific Plate during the past few million years have compressed, fractured, and uplifted the area. This tectonic deformation has buckled the lithosphere upward to create the high-standing coastal and transverse mountain ranges and downward to form the lower-lying Central Valley, Los Angeles Basin, and Ventura Basin.

> **1. g.** *Students know* how to determine the epicenter of an earthquake and know that the effects of an earthquake on any region vary, depending on the size of the earthquake, the distance of the region from the epicenter, the local geology, and the type of construction in the region.

An epicenter is that point on Earth's surface directly above the place of an earthquake's first movement, or *focus*. It is located by seismic data recorded at a minimum of three seismograph stations. The method of locating the epicenter is based on the speed of seismic waves that travel through the ground—seen as their relative times of arrival at seismic stations. These vibrations are called *P-* and *S-*waves. *P-*, or *primary*, waves are compressional with particle motion in the same direction as the wave propagation. *S-*, or *secondary*, waves are shearing with particle motion perpendicular to the direction of wave propagation. *S*-waves travel at about 60 percent of the speed of *P*-waves.

The motion of *P-* and *S*-waves and the difference in their respective velocities are easily modeled with a long and flexible spring (typically sold as a toy). Compressing and releasing a few coils at the end of the spring stretched between two students generates visible *P*-waves that travel the length of the spring and back in the opposite direction. Slapping the side of the spring generates *S*-waves that travel as sideways displacements down the length of the spring. If students measure the distance between the ends of the spring and the time it takes for the *P-* and *S*-waves to travel the full length, they can calculate the different velocities of the two waves along the spring.

Seismographs record the arrival time of the *P-* and *S*-waves. The knowledge that both waves started at the same time allows one to determine the distance of the epicenter. If three or more seismographs record distances to the same event, the epicenter of the earthquake may be determined by triangulation.

Students in grade six can be taught to locate epicenters if they are given the arrival times at various locations on a map, along with a simple velocity model. They may also be asked to locate major geologic features in and near California, such as Mount Lassen, the San Andreas Fault, the Sierra Nevada ranges, Death Valley, the Baja Peninsula, and the San Francisco Peninsula. They can draw and relate these to a map that includes outlines of the major tectonic plates. California's population is primarily concentrated near the San Andreas Fault and the system of faults that surround it. Students can plot on a map the location of their school and nearby active faults and research records of earthquake activity, ground motion, and fracturing.

Seismograph stations also record the amplitude of the ground motion, which can be used to calculate the magnitude of an earthquake (a relative measure of the amount of energy released). Magnitude is often reported according to the Richter scale, with values that generally range from around 0 to a little less than 9. Each increase of one number in magnitude represents a tenfold increase in ground shaking. Geologists may also investigate the effects of an earthquake on structures (and people's reactions) and assign an intensity value to that earthquake. The intensity values are then plotted on a map to give a more complete picture of the earthquake's effects. There are several different intensity scales, but the one most widely used in this country is the Modified Mercalli scale. This scale ranges from I (not felt) to XII (damage nearly total).

The magnitude of an earthquake is determined by the buildup of elastic strain (stored energy) in the crust at the place where ruptures (faults) may eventually occur. Unfortunately, many small earthquakes combined can release only a small fraction of the stored energy. For example, it might take as many as one million earthquakes, each at a magnitude of 4.0 on the Richter scale to release the same amount of energy as a single earthquake at 8.0. Although each increase of one number on the Richter scale reflects a tenfold increase in ground shaking, it represents nearly a thirtyfold increase in energy released. Therefore there is always a possibility that a large, destructive earthquake will release most of the stored energy. Because the materials through which earthquakes move can absorb energy and because the energy is spread over a wider area as its waves propagate outward, an earthquake tends to weaken with increasing distance from its epicenter. As earthquake waves pass through the ground, unconsolidated materials, such as loose sediments or fill, tend to shake more violently or undergo liquefaction more easily than do harder materials. Buildings made of brittle materials (e.g., reinforced concrete, brick, or adobe) tend to suffer greater earthquake damage than do those made of more flexible materials (wood). Taller buildings are often more susceptible to earthquake damage than are single story-buildings.

STANDARD SET 2. Shaping Earth's Surface

Over long periods of time, many changes have occurred in Earth's surface features. Forces related to plate tectonics have elevated mountains. Atmospheric constituents (mostly water, oxygen, and carbon dioxide) have interacted with minerals and rocks at Earth's surface, weakening them and breaking them down through a process called *chemical weathering*. Physical processes involving, for example, the growth of plants, the release of pressure as overlying material is eroded, and the repeated freezing and thawing of water in cracks, have also helped to break down rocks. Fragments are transported downslope by wind, water, and ice. Gravity by itself moves material by way of landslides and slumps (called *mass wasting*). The ultimate destination of most of the products of weathering is the ocean. These products arrive in the form of marine sediment deposits. In time the mountains are laid low, the rivers change their courses and disappear, and lakes and seas expand or dry up. Eventually sediments, which have found their way to the oceans along continental margins, are compacted and changed to rock, then uplifted by continental collision or subducted and melted under the crust. New mountains are formed, and the cycle (called the *geologic cycle*) begins anew. Each cycle takes tens of millions of years.

Chapter 4
The Science Content Standards for Grades Six Through Eight

Grade Six
Focus on Earth Sciences

> **2. Topography is reshaped by the weathering of rock and soil and by the transportation and deposition of sediment. As a basis for understanding this concept:**
>
> **a.** *Students know* water running downhill is the dominant process in shaping the landscape, including California's landscape.

Water contributes to two processes that help shape the landscape—the breakdown of rock into smaller pieces by mechanical and chemical weathering and the removal of rock and soil by erosion. Water is the primary agent in shaping California's landscape. Surface water flow, glaciers, wind, and ocean waves have all been and continue to be active throughout California and the rest of the world in shaping landscapes.

A "stream table" may be used to demonstrate the effectiveness of running water as an erosion agent. Stream tables can be easily made from plastic bins or dishpans filled with sand or gravel. The water source may be a hose, a siphon that draws from a cup, or even a drip system. Students may use either gradient or water flow rate as the independent variable. Rates of settling of different sizes of sediment through water may be demonstrated through the use of a sediment jar.

> **2. b.** *Students know* rivers and streams are dynamic systems that erode, transport sediment, change course, and flood their banks in natural and recurring patterns.

The energy of flowing water is great enough to pick up and carry sediment, thereby lowering mountains and cutting valleys. Sediment carried by a stream may be directed against solid rock with such force that it will cut or abrade the rock. The steeper the slope and the greater the volume-flow of water, the more energy the stream has to erode the land. The flow of water usually varies seasonally. At times of heavy rainfall in a watershed, a stream may flood and overflow its banks as the volume of water exceeds its containment capacity. Flooding may cause a stream to change its path. A stream bank, which consists of sediment or bedrock, may collapse and change the water's course. One example of this is a stream's tendency to shorten its length by forming oxbow lakes. This redirection of the stream's course usually takes place in natural and recurring patterns year after year.

> **2. c.** *Students know* beaches are dynamic systems in which the sand is supplied by rivers and moved along the coast by the action of waves.

The final destination of sediment is usually the ocean. Coarse sediment (sand size and larger) frequently is temporarily trapped along the shore as beach deposits while the finest sediments are often washed directly out to sea and, in some cases, carried by ocean currents for many miles. Waves that break at oblique angles to the shore move sediment along the coast. Waves wash the sand parallel to the direction in which they break, but the return water-flow brings sand directly down the slope of the beach, resulting in a zigzag movement of the sand. Students can observe differences in sand (e.g., size, color, shape, and composition) by using sand collections that may be obtained from various sources, including family and friends. The differences result from the variety of rock sources from which the sand has come, the weathering processes to which the rock has been subjected, and the completeness of the weathering (i.e., how long the rock has been subjected to weathering).

Students should attempt to identify any minerals or rocks that would indicate the kinds of weathered materials contained in the sand. Students examining sand from California beaches will find constituents such as quartz, feldspar, shell fragments, and magnetite. Magnetite is fun to extract by passing a magnet, wrapped in a plastic bag, through the sand. Magnetite may be saved and later used in place of iron filings to demonstrate magnetic field geometry for another standard.

> **2. d.** *Students know* earthquakes, volcanic eruptions, landslides, and floods change human and wildlife habitats.

Earthquakes can collapse structures, start fires, generate damaging *tsunamis,* and trigger landslides. Landslides can destroy habitats by carrying away plants and animals or by burying a habitat. Volcanic eruptions can bury habitats under lava or volcanic ash, ignite fires, and harm air quality with hot toxic gases. Floods can bury or wash away habitats.

Lives may be lost and property damaged when humans get in the way of those powerful natural processes. Although construction (and human habitation) in areas prone to natural disasters is often impossible to avoid, understanding the likelihood of such disasters and taking steps to mitigate the potential effects would be wise. Moreover, no construction too close to known hazards (e.g., on floodplains) would be advisable. Certainly, the frequency (probability) and severity of flooding, landslides, and earthquakes must be considered when one decides on land use. Making those decisions should be done after consideration of many factors, including the use of scientific evidence to predict catastrophic events and the local impacts. Although catastrophic events are usually adverse in the short term, some of them may be beneficial in the long term. For example, river floods may deliver new, nutrient-enriched soil for agriculture. Other catastrophic changes may introduce new habitats, allow fresh minerals to surface, change climates, or give rise to new species.

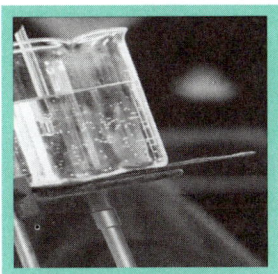

STANDARD SET 3. Heat (Thermal Energy) (Physical Science)

Prior to the nineteenth century, the transfer of heat was assumed to be due to the flow of a substance called *caloric,* an invisible, weightless fluid whose total quantity remained constant. The caloric theory subscribed to the belief that an object became hot when it was permeated by a large quantity of caloric and cooled when some of its caloric flowed into other objects that had less caloric. This model was upset by the work of two scientists: Benjamin Thompson (later known as Count Rumford) and James Joule. Rumford supervised the boring of cannons. He noted that the water kept in the bores to prevent overheating boiled continuously. This boiling was supposedly caused by the caloric that flowed from the metal of the cannon as it was cut.

From his observations, however, Rumford deduced that this explanation could not be correct because the boiling continued even when the boring tool became so dull that it no longer had any effect on the metal. Apparently, the caloric was being produced out of nothing. Rumford concluded that it was the work needed to turn the dull tool, instead of caloric transfer, that was being converted into heat. In a series of experiments, Joule showed that a given amount of mechanical work always produced the same amount of heat no matter what kind of mechanical work was done. This demonstration established that heat is indeed a form of energy. Today, it is known that heat is energy contained in the random motion of atoms and molecules and that to heat an object is to increase the energy so stored.

Although students will not be exposed to kinetic molecular theory until high school, teachers who understand the following points will be better able to discuss the subjects of heat and heat transfer. The transfer of heat from a warmer object to a colder object is referred to as *heat flow.* Heat may be transferred by conduction, convection, or radiation. Standard Sets 3 and 4 in grade six deal in depth with the relationships between heat and convection in Earth's mantle, oceans, and atmosphere. Material covered in those standards will build a foundation for the study of heat. Students will learn that atoms are free to move in different ways in solids, liquids, and gases and that heat may be given off or absorbed during chemical reactions. The concept that heat is a form of energy associated with the motion of atoms and molecules is covered in high school. Students in grade six will study the relationship between work and heat flow and will be required to solve problems related to this subject.

> **3. Heat moves in a predictable flow from warmer objects to cooler objects until all the objects are at the same temperature. As a basis for understanding this concept:**
>
> **a.** *Students know* energy can be carried from one place to another by heat flow or by waves, including water, light and sound waves, or by moving objects.

Energy is transferred from one object to another as the result of a difference in temperature. *Heat flow* is the transfer of energy from a warmer object to a cooler object.

A wave is an oscillating disturbance that carries energy from one place to another without a net movement of matter. For example, sound waves from one vibrating object can cause other objects, such as eardrums, to vibrate. Electromagnetic waves can also carry energy. One example of this phenomenon is the transfer of heat from the Sun to Earth. Students may think of the infrared radiation escaping from a bed of hot coals and warming their hands as another example of heat flow.

Energy can also be transferred by the movement of matter. For example, the energy supplied by the pitcher's arm transports a pitched baseball to the catcher's mitt.

> **3. b.** *Students know* that when fuel is consumed, most of the energy released becomes heat energy.

When fuel is burned, energy stored in the fuel's chemical bonds is released as heat and light. Only a small portion of the energy contained in the original fuel remains locked in the waste products left over after the fuel has been consumed. Although the heat derived from fuels is often used in turn to drive engines that perform useful work, an important understanding is that even the work performed ultimately tends to be transformed into heat. For example, an automobile set into motion and braked to rest transforms most of its kinetic energy into heating the brake pads by friction. As a demonstration the teacher might light a candle in the classroom and let students know that the wax in the candle is the fuel that combines with oxygen in the air to produce both heat and light. Most of the heat is transferred to the room by the hot gases rising from the flame. Glowing particles of soot (the source of the yellow light) also transfer energy from the flame. Students might be asked to develop an explanation of how heat is transferred from the burning fuel to a container of water heated by the candle, using the concepts and principles called for in this standard set.

> **3. c.** *Students know* heat flows in solids by conduction (which involves no flow of matter) and in fluids by conduction and by convection (which involves flow of matter).

This standard focuses on differences between heat transfer by conduction and by convection and begins to build an understanding of the kinetic molecular theory of heat transfer. In both solids and fluids (liquids and gases), heat transfer is measured by changes in temperature.

Conduction occurs when a group of atoms or molecules whose average kinetic energy is greater than that of another group transfers some of that excess energy by means of collisions. Because hot objects have atoms with greater average kinetic energy than do cold ones, there is a transfer of this kinetic energy from hot to cold. In a solid the atoms vibrate in place, but energy may still be transferred from atom to atom as happens when a pan is placed on a stove and its handle becomes hot. The same mechanism describes the conduction of heat in liquids and gases, where the atoms are free to slip past one another provided there is no cumulative flow in the material. To demonstrate conduction a teacher might wrap some paper (to form a handle) around the end of a metal rod about 30 centimeters long and use paraffin to attach a series of thumbtacks, spaced about two centimeters apart, along the rod. The teacher then holds the rod by the handle and places the free end over a candle or in a burner flame. As heat is conducted along the rod, the tacks drop away one by one.

Convection occurs because most fluids become less dense when heated; the hot fluid will rise through cold fluid because of the hot fluid's greater buoyancy. As hot fluid arises away from a heat source, it may cool, become denser, and sink back to the source to be warmed again. The resulting circulation is called a *convection current.* Convection currents account for the water in a kettle reaching a uniform temperature although the kettle is warmed only at the bottom. The effects of convection may be investigated by placing finely shredded paper into a large heat-resistant beaker or roasting pan filled with cold water. After the paper is saturated and sinks to the bottom of the container, the teacher may apply heat from a hot plate and note that the paper particles move upward near the heat source and downward away from it.

> **3. d.** *Students know* heat energy is also transferred between objects by radiation (radiation can travel through space).

Another form of energy transfer between objects is radiation: the emission and absorption of electromagnetic waves. Radiation is fundamentally different from conduction and convection in that the objects do not have to be in contact with each other or be joined by a solid or fluid material. Heating by sunlight is an obvious example of radiant energy transfer. Both the heat and the light that can be seen are forms of electromagnetic radiation. Calling attention to this fact may help dispel the common misconception that all radiation is harmful.

STANDARD SET 4. Energy in the Earth System

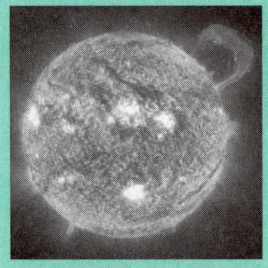

Energy that reaches Earth's surface comes primarily as radiation from the Sun. Solar energy includes the full electromagnetic spectrum, but most of it is carried in the visible region. Because the atmosphere is transparent to visible light, most of this incoming energy is transferred to Earth's surface. Conductive transfer and reradiation of this energy heat the lower atmosphere and result in convection currents that distribute the heat into the atmosphere.

Solar radiation heats Earth's surface unevenly, resulting in thermal gradients in the atmosphere. Variations in the angle of sunlight influence the amount of energy reaching each square meter of Earth's surface and largely account for the uneven heating of the surface. The angle of sunlight varies because of Earth's spherical shape and because the Sun's rays travel in a straight line parallel to one another. If a surface area of this planet is directly perpendicular to the Sun (meaning the Sun is directly overhead), then the rays strike at a 90-degree angle, resulting in maximum absorption of solar radiation because the energy is concentrated on a relatively small area. As the surface curves away from this spot, the angle at which sunlight strikes it becomes smaller, and the same amount of solar radiation is spread over a broader area.

The uneven heating of Earth's surface and the tilt of its axis (66.5 degrees to the orbital plane or 23.5 degrees to the perpendicular) account for the seasons and extremely cold north and south poles. Clouds and the varied reflectivity of Earth's surface contribute to uneven heating. In general, however, the total solar energy transferred to Earth is nearly constant, and all the energy gains and losses are in balance. Consequently, Earth enjoys climates that are relatively stable for thousands of years, with predictable temperature ranges and weather patterns that can be broadly forecast.

Various heat exchange mechanisms operate in the Earth system. Ocean surface water is heated by the Sun and mixed by convection currents. The atmosphere exchanges heat with the oceans and land masses by means of conduction. Warm air near Earth's surface rises and cooler air descends, causing atmospheric convection currents. Different parts of the ocean have different temperatures and salinities, resulting in deep convection currents. The convection currents in the atmosphere move evaporated water away from ocean surfaces; from there the water vapor can be picked up by winds and carried to other locations where it may condense as precipitation. In this manner both heat and water are transported.

The observed patterns of surface winds are mostly the result of convection currents caused by uneven surface heating. Winds are deflected by the Coriolis effect (caused by the west-to-east turning of Earth) and by topography. Latitude, winds (speed, direction, and moisture content), and the elevation of the land and its proximity to the ocean largely determine the climate and corresponding weather patterns in any particular region.

Earth's crust contains localized concentrations of internal heat, as evidenced by volcanoes, hot springs, and geysers. However, the total amount of heat transferred to the atmosphere from Earth's crust is minute compared with the amount of heat the surface receives from the Sun.

> **4. Many phenomena on Earth's surface are affected by the transfer of energy through radiation and convection currents. As a basis for understanding this concept:**
>
> **a.** *Students know* the sun is the major source of energy for phenomena on Earth's surface; it powers winds, ocean currents, and the water cycle.

Radiation from the Sun penetrates the atmosphere by heating the air, the oceans, and the land. Solar radiation is also converted directly to stored energy in plants through photosynthesis. The Sun is a constant, close-to-uniform source of energy that is responsible for the climate and weather, drives the water cycle, and makes life possible on Earth.

> **4. b.** *Students know* solar energy reaches Earth through radiation, mostly in the form of visible light.

A full-wavelength spectrum of electromagnetic energy is present in solar radiation from below the infrared to above the ultraviolet. However, most of the energy radiated by the Sun is in the visible or near visible part of the light spectrum, and that is largely the part that penetrates the transparent atmosphere and reaches Earth's surface. Because blue light is scattered by the atmosphere more than yellow light, the sky looks blue and the Sun looks yellow. Students should understand that both long- and short-wavelength radiation may interact in various ways with atmospheric constituents and may be absorbed by atmospheric constituents in different amounts; however, the wavelengths of visible light are not greatly absorbed by any atmospheric constituent.

> **4. c.** *Students know* heat from Earth's interior reaches the surface primarily through convection.

Heat from the interior of Earth moves toward the cooler crustal surface. Rock is a poor conductor of heat; therefore, most of the transfer of heat occurs through convection. Convection currents in the mantle provide the power for plate tectonic movements. Heat reaching Earth's surface in this manner is transferred to the atmosphere in relatively small amounts.

> **4. d.** *Students know* convection currents distribute heat in the atmosphere and oceans.

Convection plays a central role in transferring heat energy from place to place in the atmosphere and ocean. Uneven heating of the land and ocean causes convection currents. This movement of air and water creates the wind and ocean currents that are deflected by the geography of the land and the rotation of Earth. Students can investigate atmospheric convection currents on a small scale by using a smoke chimney or fog chamber. In the absence of more sophisticated equipment, much can be observed about atmospheric convection by studying what happens to visible water droplets (condensing steam) as they exit a boiling teakettle.

There are several ways to investigate convection currents in a liquid. One way is to float a large ice cube (tinted with food coloring) on hot water and trace the resulting convection currents. Another way is to heat one end of an elongated cake pan full of water. Convection may be observed by adding drops of food coloring.

> **4. e.** *Students know* differences in pressure, heat, air movement, and humidity result in changes of weather.

Changes in local temperatures, atmospheric pressure, wind, and humidity create the weather that everyone experiences. All those effects are connected directly to the processes associated with the transfer of solar energy to Earth and redistribution of that energy in the form of heat. Precipitation occurs when moist air is cooled below its condensation temperature (dew point).

Great currents circle the globe in the convecting atmosphere and ocean, created by atmospheric pressure and temperature gradients that, in turn, spin off local winds and eddies. Temperature differences also lead to changes in humidity and precipitation. The local set of these descriptive measures is called weather, and the changes result in weather patterns. The long-term seasonal average of these weather patterns defines the climate of an area.

STANDARD SET 5. Ecology (Life Sciences)

All living organisms are a part of dynamic systems that continually exchange energy. These systems are regulated by both biotic and abiotic factors. Nutrients needed to sustain life in an ecosystem are cycled and reused, but the energy that flows through the ecosystem is lost as heat and must constantly be renewed. Green plants are the foundation of the energy flow in most ecosystems because they are capable of producing their own food by photosynthesis. Because energy is either used by consumers or depleted in a logical progression, it can be said to flow through a food web (also known as a food chain). A food web may be represented as an energy pyramid with green plants as a base, midlevel consumers in the middle, and a few top-level predators at the apex. Scavengers and decomposers are the final members of an energy pyramid as they clean up the environment and return matter (nutrients) to Earth. A food web can also show the various roles played by plants and animals as producers, consumers, and decomposers.

> **5. Organisms in ecosystems exchange energy and nutrients among themselves and with the environment. As a basis for understanding this concept:**
>
> **a.** *Students know* energy entering ecosystems as sunlight is transferred by producers into chemical energy through photosynthesis and then from organism to organism through food webs.

A food web depicts how energy is passed from organism to organism. Plants and photosynthetic microorganisms, or producers, are the foundation of a successful food web because they do not need to consume other organisms to gain energy. Instead, they gain their energy by transforming solar energy through photosynthesis into chemical energy that is stored in their cells.

> **5. b.** *Students know* matter is transferred over time from one organism to others in the food web and between organisms and the physical environment.

Energy and matter are transferred from one organism to another organism through consumption. Plants are eaten by primary consumers (herbivores); most herbivores are eaten by secondary consumers (carnivores); and those consumers are eaten by tertiary consumers (often top-level predators). At the microscopic scale photosynthetic bacteria (cyanobacteria) and protists or single-celled eukaryotic organisms (e.g., dinoflagellates) are consumed by heterotrophic protists (e.g., amoebae and ciliates), which are also called *protozoans*. Protozoans are consumed by other larger protozoans and by small animals such as cnidarians, arthropods, and nematodes. Energy is transferred from organisms (microorganisms, plants, fungi, and animals) to the physical environment through heat loss. Carbon is returned to the physical environment as airborne carbon dioxide through the respiration of organisms. Water is also cycled. Students may use science texts and other library materials to research organisms included in the food webs of particular ecosystems. Students can draw model food webs to demonstrate how food energy is transferred from plants to consumers and from consumer to consumer through predation. Students can also depict the hierarchy of consumers and the transfer and loss of energy from herbivores through secondary consumers to the top carnivores in a food web or energy pyramid. Students should know that energy is lost to the physical environment at every hierarchical level.

> **5. c.** *Students know* populations of organisms can be categorized by the functions they serve in an ecosystem.

Organisms in a population may be categorized by whether they are producers of chemical energy from solar energy (e.g., plants and photosynthetic microorganisms) or consumers of chemical energy (e.g., animals, fungi, and heterotrophic protists) and, if they are consumers, whether they are predators, scavengers, or decomposers. Many consumers may be categorized in multiple ways, such as omnivores that eat both plants and animals and opportunistic consumers that act as both predators and scavengers. Teachers may provide the class with a nonordered, noncategorized list of four or five plants, eight to ten consumers (four or five primary consumers, three or four secondary consumers, and one or two tertiary consumers [or top-level predators]), one or two decomposers, and one or two scavengers. Using a science text or appropriate research materials from the school library, students can identify the organisms by food web order and ecological function. Students can then arrange the organisms into an energy pyramid with the decomposers and scavengers identified and noted separately. The final task is to draw arrows between members of the pyramid to depict the predation sequence.

> **5. d.** *Students know* different kinds of organisms may play similar ecological roles in similar biomes.

Ecological roles are defined by the environment and not by any particular organism. For example, Australia has plants that are unique to that continent yet play the same role as other kinds of plants in similar environments elsewhere. In the rain forests of South America, the mammalian consumers and predators are placental (nonmarsupial) sloths, deer, monkeys, rodents, and cats. In the rain forests of Australia, marsupial kangaroos, wallabies, bandicoots, and so forth play the same ecological roles. Students may be assigned or may choose to research specific organisms that occupy similar biomes in widely separated geographic locales. Students should be encouraged to use a variety of library resources, such as expository texts, the Internet, CD-ROM reference materials, videos, laser programs, or periodicals.

5. e. *Students know* the number and types of organisms an ecosystem can support depends on the resources available and on abiotic factors, such as quantities of light and water, a range of temperatures, and soil composition.

There is a greater variety of types of organisms in temperate or tropical environments than in deserts or polar tundra. The number of organisms supported by an ecosystem also varies from season to season. More organisms thrive during temperate summers than can survive icy winters. More organisms can multiply during a desert's cooler, wetter winters than can live through its hotter, drier summers. Students should understand that the richness of plant growth controls the diversity of life types and number of organisms that can be supported in an ecosystem (the base of the pyramid). Richness of plant growth depends on abiotic factors, such as water, sunlight, moderate temperatures, temperature ranges, and composition of the soils. To support vigorous growth, soils must contain sufficient minerals (e.g., nitrogen, phosphorus, potassium) and humus (decomposed organic materials) without excess acidity or alkalinity. The teacher may point out that the number of plant-eating animals in an ecosystem depends directly on the available edible plants, and the number of predators in a system depends on the available prey.

STANDARD SET 6. Resources

Although this standard set deals with the concept of finite resources, the emphasis is on energy. Much of the energy used worldwide is derived from *nonrenewable* fossil fuels, such as coal, oil, and natural gas. Those resources are being consumed at rates far faster than their geologically slow formation rates. Uranium (for fission energy) and deuterium (for fusion energy) are also finite but are in abundant supply (deuterium is almost inexhaustible). Industrial waste and pollution result from nuclear power generation and the burning of fossil fuels. The extraction (mining) and processing (smelting) of both energy and nonenergy resources also have environmental consequences. There are numerous types of renewable energy resources, including solar, wind, hydroelectric, and geothermal, but they are largely undeveloped or underdeveloped. Knowing the forms, conversion processes, end-uses, and impact of wastes involved in using natural resources, whether for energy or materials, is critical in making decisions and trade-offs about how those resources will be used.

> **6. Sources of energy and materials differ in amounts, distribution, usefulness, and the time required for their formation. As a basis for understanding this concept:**
>
> **a.** *Students know* the utility of energy sources is determined by factors that are involved in converting these sources to useful forms and the consequences of the conversion process.

Useful energy sources are those that can be converted readily to forms of energy needed for heat, light, and transportation. Technologies have been developed to convert various forms of energy (e.g., oil, gas, solar, nuclear, wind, and wave) to meet those needs. For example, manufacturers have learned how to refine oil to make gasoline, which can then be used in the combustion engine to provide transportation or in power generators to produce electricity. The energy sources considered the most useful are those for which the most cost-effective conversion technologies have been developed. For transportation purposes solar energy is not considered as useful mainly because inexpensive and efficient solar energy storage systems have yet to be developed. Until those systems are developed, solar energy will not be able to meet the demand for reliable levels of power or provide a driving range comparable to that provided by gasoline and diesel fuels.

Students should be taught the concept of nonmonetary costs of energy. Mining coal leaves large, open pits and may pollute the atmosphere with the exhaust of heavy mining machinery. Power plants may also pollute the atmosphere with the exhaust from burning fossil fuels. Nuclear power plants must exhaust excess heat, often in the form of hot water introduced into rivers and oceans. Hydroelectric energy, although it is renewable and has no effect on air quality, requires the damming of streams—a measure that carries upstream environmental implications and downstream consequences on sediment load and beaches as well as the possibility of disaster caused by dam failures. Students may use published materials and Internet resources (consistent with Internet-use policies in effect at the school) to research, evaluate, and report on the environmental consequences. In this way they can develop a clearer understanding of the nonmonetary costs of energy in relation to environmental protection (conservation). Students can rate the environmental advantages and disadvantages of heating a home with electricity, natural gas (or propane), solar power, oil, or coal.

> **6. b.** *Students know* different natural energy and material resources, including air, soil, rocks, minerals, petroleum, fresh water, wildlife, and forests, and know how to classify them as renewable or nonrenewable.

Renewable and nonrenewable energy and natural resources depend on both the process and the time needed to create energy sources. Solar energy cannot be exhausted nor can fuels for fusion; therefore, they are sometimes referred to as *renewable.* Hydroelectric power is dependent on the water cycle (driven by solar energy) and is considered a renewable resource. Because biomass will grow back quickly to replace that used for fuel or materials, it is also considered renewable. However, if habitats and species are lost in the process of harvesting the biomass, the resources are nonrenewable in that sense. Trees used for fuel or building materials can be replaced only if the rate of use does not exceed the time needed to grow replacement trees and if the land is not altered to become unusable for that purpose. Fossil fuels (coal, oil, natural gas) were formed on geologic time scales and are considered nonrenewable resources.

> **6. c.** *Students know* the natural origin of the materials used to make common objects.

This standard deals with the ultimate sources of common objects. Students often do not consider or even know the natural origins of commonly used goods. They must be reminded that manufactured items do not appear magically and that the ultimate cost of acquiring the objects goes far beyond the price sticker. Students can count the objects in their classroom to make an inventory and trace them back to the natural materials from which they were manufactured. Students can then classify the materials as renewable or nonrenewable. They may need to do some careful research to discover the origins of some materials. For example, a simple pencil contains wood and lead. But the pencil lead is actually a mixture of graphite and clay. If the pencil has an eraser, the rubber from a plant (or plastic from petroleum) and metal for the holder must be included. Students may realize in looking at clothing, paper, paint, tiles, windows, projectors, computers, chairs, books, chalk, crayons, brooms, and so on that plastics and synthetic materials are derived from oil.

STANDARD SET 7. Investigation and Experimentation

Students are expected to formulate a hypothesis for the first time. A hypothesis is a proposition assumed as a basis for reasoning and often subject to the testing of its validity. The scientific hypothesis provides an explanation of a set of observations and may incorporate observations, concepts, principles, and theories about the natural world. Hypotheses lead to predictions that can be tested. If the predictions are verified, the hypothesis is provisionally corroborated. If the predictions are incorrect, the original hypothesis is proved false and must be abandoned or modified.

Hypotheses may be used to build more complex inferences and explanations. Hypotheses always precede predictions. However, for simple investigations the hypothesis that led to a prediction may not be easily identified because of its simplicity or its complexity. Prediction follows observation in grades three to five. After grade six students should recognize and develop a hypothesis as a part of their experimental design. In grade six the focus on earth science can provide many opportunities in the Investigation and Experimentation standards to develop students' ability to design experiments and to select and use tools for measuring and observing.

7. **Scientific progress is made by asking meaningful questions and conducting careful investigations. As a basis for understanding this concept and addressing the content in the other three strands, students should develop their own questions and perform investigations. Students will:**

   a. Develop a hypothesis.

   b. Select and use appropriate tools and technology (including calculators, computers, balances, spring scales, microscopes, and binoculars) to perform tests, collect data, and display data.

   c. Construct appropriate graphs from data and develop qualitative statements about the relationships between variables.

   d. Communicate the steps and results from an investigation in written reports and oral presentations.

   e. Recognize whether evidence is consistent with a proposed explanation.

   f. Read a topographic map and a geologic map for evidence provided on the maps and construct and interpret a simple scale map.

   g. Interpret events by sequence and time from natural phenomena (e.g., the relative ages of rocks and intrusions).

   h. Identify changes in natural phenomena over time without manipulating the phenomena (e.g., a tree limb, a grove of trees, a stream, a hillslope).

Grade Seven — Focus on Life Sciences

Now is an exciting time for the study of life sciences. Knowledge of biological systems is expanding rapidly, and the development of new technologies has led to major advances in medicine, agriculture, and environmental management. A foundation in modern biological sciences, with an emphasis on molecular biology, is essential for students who will become public school science teachers, college and university science professors and researchers, and specialists in technological fields.

Another definitive reason for a focus on life science in grade seven is the students' own biological and behavioral transition into early adolescence. Young adolescents make decisions that may have an enormous influence on their lives. The study of life science provides a knowledge base on which adolescents can make well-informed and wise decisions about their health and behavior. The relevance of the curriculum to students' lives helps students to maintain an interest in science and to expand their knowledge of the natural sciences.

The *Health Framework for California Public Schools* is a valuable resource for science teachers.[4] It contains grade-level expectations for health education that provide important connections to the life science curriculum. Specific statutes require parental notification regarding the teaching of topics related to human growth and development.

STANDARD SET 1. Cell Biology

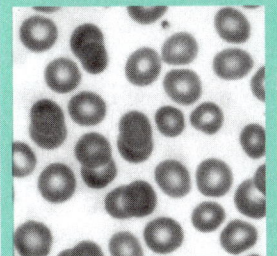

In the middle grades students expand their knowledge of living systems to include the study of cells, the fundamental units of life.

In grade five students learned about the organs or tissues for respiration, digestion, waste disposal, and transport of materials in plants and animals. They were also first introduced to cellular functions when they studied cellular respiration in animals and plants and photosynthesis in plants. These studies are complemented in grade seven by new material on the cellular organelles responsible for those functions.

The standards in grade six covered ecology, and students in that grade learned how energy in the form of sunlight is transformed by producers into chemical energy through the process of photosynthesis. The study of energy transfer through food webs provided a foundation for a more detailed exploration at the cellular level of how plant chloroplasts capture sunlight energy for photosynthesis and how mitochondria liberate energy for the work that cells do.

> **1. All living organisms are composed of cells, from just one to many trillions, whose details usually are visible only through a microscope. As a basis for understanding this concept:**
>
> **a.** *Students know* cells function similarly in all living organisms.

There are fundamental aspects of cell function that are similar regardless of the organism in which the cell resides. For example, cells contain a DNA (deoxyribonucleic acid) genome (i.e., all genetic material in a cell) and express the genome by using a universal genetic code. The biochemical pathways in cells, such as those for cell division and energy production, are strikingly similar even though the cells serve different functions in and between organisms. Many proteins synthesized by cells have similar functions, such as serving as enzymes that promote chemical reactions in the cell. There are significant functional differences between cells in an organism as they become differentiated, or specialized (e.g., a liver or a brain cell). There are also significant differences between cells in different environments, such as the *Escherichia bacterium* living in an intestine or a *Thermophilus bacterium* living in a superheated geyser. Biological science has been greatly advanced by the uncovering of both similarities and differences among cells.

> **1. b.** *Students know* the characteristics that distinguish plant cells from animal cells, including chloroplasts and cell walls.

Plant cells are surrounded by a cell wall (made primarily of cellulose) that is rigid and limits the shape of the cell membrane. Animal cells, however, are not surrounded by a cell wall, and their shape is defined by their underlying cytoskeleton. Many plant cells contain chloroplasts and a central vacuole, neither of which is found in animal cells. Those differences between plant and animal cells may be made apparent by microscopy as sections of plant and animal tissue are appropriately stained to highlight the structures. Images of cells are also available on the Internet and in textbooks. Labeled diagrams will help students learn about structures that are too small to be seen with the use of classroom microscopes.

> **1. c.** *Students know* the nucleus is the repository for genetic information in plant and animal cells.

Chromosomes containing genes reside in the nucleus. When an interphase cell is observed by using a light microscope, the inside of the nucleus may appear to be homogeneous because the chromosomal DNA is not condensed. In an appropriately fixed and stained section of onion root (obtainable from commercial sources), the DNA will be visible as a disk-shaped area, apparently constrained within a nucleus. This is the best stage in which to visualize DNA in learning the content of the standard. If the root tissue had a high rate of growth at the time it was sectioned and fixed, a fraction of the cells may be in one of the stages of mitosis. In that case the chromosomes will be visibly condensed but will not be limited by a nuclear membrane. This phenomenon must be explained carefully so that students do not develop a misconception about the distribution of DNA in a cell on the sole basis of their observation of mitotic chromosomes.

> **1. d.** *Students know* that mitochondria liberate energy for the work that cells do and that chloroplasts capture sunlight energy for photosynthesis.

Students may already understand that the food they eat provides them with energy in an informal sense. At the cellular level the mitochondrion is responsible for efficiently extracting the chemical energy from molecules that have been broken down mostly from ingested food. The energy liberated by mitochondria is still stored in the form of chemical energy but in molecules that are readily accessible for energy release. Chloroplasts use pigments to absorb the energy in sunlight. This captured energy is used to drive a chemical reaction within the chloroplast in which carbon dioxide from the air is used as a source of carbon to form sugar molecules from which mitochondria extract energy used in the cell.

> **1. e.** *Students know* cells divide to increase their numbers through a process of mitosis, which results in two daughter cells with identical sets of chromosomes.

Just as living organisms are said to have a life cycle that relates to their periods of growth and reproduction, cells are said to have a "cell cycle." Cells reproduce themselves by a process called *mitosis*. The process takes place after a period of growth during which the DNA in the nucleus is replicated and cytoplasmic organelles, such as mitochondria and chloroplasts, are doubled in number. During mitosis the replicated DNA chromosomes are segregated so that each daughter cell receives exactly the same number of chromosomes of each type (e.g., two of each type in a diploid organism). Students may observe mitotic chromosomes by light microscopy in a stained section of rapidly growing tissue. Time-lapse videos and movies of cell division will also help to illuminate the process of chromosome segregation.

> **1. f.** *Students know* that as multicellular organisms develop, their cells differentiate.

In most multicellular organisms there is a division of labor among cells. Some cells in humans are brain cells; others are stomach, skin, or muscle cells. Although those cells are clearly different, their ancestry can be traced back to a single fertilized egg. During the development of an embryo, some cells become fixed in their developmental program and are said to be *differentiated*. For example, cells that will eventually divide to give rise to the stomach and intestines are distinguished at a very early stage from cells that will divide to give rise to the central nervous system and eyes. At later stages of development, a more fine-grained differentiation takes place. For example, some cells in the retina of the eye become rod cells (for vision in dim light) and others become cone cells (for color vision). After differentiation, most cells in humans lose the ability to become other types of cells.

In plants the cells often retain the ability to differentiate into other tissues. For example, a leaf of an African violet can set roots in soil and develop into a new plant. Although the leaf is clearly differentiated, it is not fixed in its developmental potential in the way that animal cells typically are (an exception being the animal's germ cells that lead to eggs and sperm).

STANDARD SET 2. Genetics

Genetics is the study of the biological processes involved in transmitting the unique characteristics of an organism to its offspring. Discovering the genetic principles and mechanisms that account for growth, senescence, and heredity has been a great accomplishment of modern science. Gregor Mendel's studies of pea plants revealed the concept of genes and the rules for the inheritance of traits. Today it is understood that those rules are based on the chemical composition and structure of DNA. Students in grade seven will learn some of those rules, which will serve as a foundation for high school biological sciences.

> **2. A typical cell of any organism contains genetic instructions that specify its traits. Those traits may be modified by environmental influences. As a basis for understanding this concept:**
>
> **a.** *Students know* the differences between the life cycles and reproduction methods of sexual and asexual organisms.

Sexual reproduction entails fertilization, an event in animals that requires the fusion of an egg cell with a sperm cell. The fertilized egg (the zygote) goes through a series of cell divisions (mitosis) and developmental steps to generate a new organism genetically related to its parents. Pollination of flowering plants and growth of a new genetically related plant from seed should also be presented as examples of a sexual life cycle.

Some organisms exclusively reproduce without a fertilization event. This method is called *asexual* reproduction. Protists (single-celled eukaryotic organisms) often have no known sexual cycle and reproduce solely by mitotic division. Fungi and plants often have both sexual and asexual methods of reproduction. For example, plants may be propagated from a seed (sexual method) or a cutting (asexual method). Although a seed is related to two parental plants, a cutting is genetically identical to the plant from which it was taken. Some primitive animals, such as the flatworm Planaria, can divide themselves asexually into two genetically identical organisms. Asexual reproduction should not be confused with reproduction in primitive animals such as nematodes. Asexual reproduction should also not be confused with hermaphroditic sexual reproduction that entails fusion of eggs and sperm generated by a single organism.

> **2. b.** *Students know* sexual reproduction produces offspring that inherit half their genes from each parent.

Sexual reproduction combines the genetic material from two different cells. In most animal species, including humans, the genetic information is contributed from two different parents, nearly half from the biological mother and half from the biological father. Mitochondria DNA is derived solely from the mother, making possible the tracing of heritage from grandmothers to grandchildren with great certainty. During fertilization the egg and sperm cells combine their single sets of chromosomes to form a zygote containing two sets, or the diploid number, of chromosomes for a species (half from each parent).

> **2. c.** *Students know* an inherited trait can be determined by one or more genes.

In the preceding standard the idea of genes was introduced to students as something inherited from each parent in roughly equal quantities. This standard draws a correlation between genes and the inherited traits or features of an organism. For example, attached or unattached earlobes is an inherited trait typically determined by a single gene (inherited from each parent). Having attached or unattached earlobes is very likely just one visible manifestation of that particular gene, which may have many other important roles during development that have not been cataloged. A single gene may affect more than one trait or feature in an organism. Many traits, such as hair and eye color, are determined by multiple genes and do not have simple patterns of inheritance. Although an organism's genes define every inherited trait, there is not always a one-to-one correspondence between trait and gene.

> **2. d.** *Students know* plant and animal cells contain many thousands of different genes and typically have two copies of every gene. The two copies (or alleles) of the gene may or may not be identical, and one may be dominant in determining the phenotype while the other is recessive.

This standard introduces some principles of Mendelian genetics. The most significant concept is that genes exist in multiple versions, called *alleles,* and these units of heredity are not typically changed during mating. Prior to acceptance of Mendel's laws, people believed that the mixing of genetic information was similar to mixing paint; the information (like red or white paint) could be blended to form a combined version (like pink paint) that could be blended still further (making it more white or more red). Using true-breeding strains of peas with variation of a single gene (such as flower color), Mendel showed that this model of blending was incorrect.

In grade seven students will learn that every person has tens of thousands of genes and that there are slight variations, or alleles, of these genes in every individual. Using the correct vocabulary is important: A person with a genetic disorder does not have the *gene* for that trait, but it might be said that the person has the genetic *allele* for that trait. Every person has every gene (and usually in two copies), but some people have an abnormal or different version (or versions) that can lead to a disorder or different trait. The genetic traits of an individual are determined by which alleles of genes are inherited from each parent and how those alleles work together. Some alleles are dominant, meaning that they overcome the influence of the other (recessive) alleles. In grade seven students learn to interpret the genotype-phenotype relationship in offspring (for example, on a premade Punnett Square diagram). In high school biology students will learn many of the details of genetics. Therefore, because this standard provides a foundation for transmission genetics in high school biology, the details of genetics (including the construction of the Punnett Square model) may be deferred.

> **2. e.** *Students know* DNA (deoxyribonucleic acid) is the genetic material of living organisms and is located in the chromosomes of each cell.

Chromosomes in eukaryotes are complexes of DNA and protein. Chromosomes organize the genetic make-up of a cell into discrete units. Humans, for example, have 23 pairs of chromosomes that vary in size. When looking through a microscope at an appropriately stained section of onion root tip, students may see cells that are engaged in mitosis and that have visible, condensed chromosome structures. The proteins in a chromosome help to support its structure and function, but the genetic information of a cell is uniquely stored in the DNA component of the chromosome.

STANDARD SET 3. Evolution

In grade two students developed simple notions of inheritance and variation within a species. Those notions are foundational for the study of evolution, as are the studies in grade three of adaptations to an environment and the processes of extinction. Charles Darwin was a naturalist who traveled widely; students in grade three are retracing his steps when they develop their knowledge of organisms in a wide variety of earth biomes. Students in grade four learn about the survivability (or fitness) of plants and animals in an environment, and students in grade six are provided with a background in earth science. The standards in this set provide a foundation for learning about natural selection in grade seven and understanding the fossil record to be a line of evidence for the evolution of plants and animals.

> **3. Biological evolution accounts for the diversity of species developed through gradual processes over many generations. As a basis for understanding this concept:**
>
> **a.** *Students know* both genetic variation and environmental factors are causes of evolution and diversity of organisms.

In grade two students learned that some characteristics of an organism are inherited from the parents and that some are caused or influenced by the environment. They also learned that there is variation among individuals in a population. This standard takes these simple ideas to much greater depth by explicitly referring to *environmental factors* and *genetic variation*. Environmental factors are a cause of natural selection, but as the term *selection* implies, there must also be favorable and unfavorable traits uncovered in the population. *Genetic* variability must precede natural selection, or there is some risk that no individuals in the population will survive a crisis. This principle is evident in the worldwide cheetah population and in other endangered species with much genetic homogeneity. Having little genetic variation to spread the risk makes a population more susceptible to extinction, for example, by succumbing to an infectious disease for which there is no natural resistance.

> **3. b.** *Students know* the reasoning used by Charles Darwin in reaching his conclusion that natural selection is the mechanism of evolution.

In his book *On the Origin of Species by Means of Natural Selection,* Charles Darwin explained his line of reasoning for natural selection as the primary mechanism for evolution.[5] Darwin proposed that differences between offspring would occur randomly. Some of those differences would be hereditary and affect an individual offspring's ability to survive and reproduce within a particular environment and ecological setting. With the passage of succeeding generations, those individuals best suited to particular environments would tend to have more progeny and those less well suited would have fewer progeny or even become extinct. Darwin called this process *natural selection* because environmental and ecological conditions essentially "select" certain characteristics of plants and animals for survival and reproduction. Darwin proposed that over very long periods of time, natural selection acting on different individuals within a population of organisms might account for all the great varieties of species seen today and for the great number of extinct and nonextinct species found in the fossil record. Darwin's proposal that natural selection is the mechanism for evolution was drawn in part from the ideas of Thomas Malthus' *Essay on the Principle of Population.*[6] Malthus presented his argument that human populations have a tendency to grow faster than their food supply, causing shortages and a "struggle for existence." Darwin's observations in the Galapagos Islands led him to think that this "struggle for existence" might be generalized to animals and plants.

Chapter 4
The Science Content Standards for Grades Six Through Eight

Grade Seven
Focus on Life Sciences

> **3. c.** *Students know* how independent lines of evidence from geology, fossils, and comparative anatomy provide the bases for the theory of evolution.

Independent lines of evidence from geology, the fossil record, molecular biology, and studies of comparative anatomy support the theory of evolution. Many decades before Darwin proposed his theory, geologists knew that sedimentary rocks formed an important history of life on Earth. Geologically younger rock layers are usually near the top, and older layers are successively closer to the bottom of sedimentary formations. Sometimes the normal sequence of sedimentary layers has been overturned by tectonically caused folding and faulting, resulting in older rock units resting on top of younger units.

Some of the organisms that lived in or were buried by the original sediment were preserved as fossils while the sediment hardened into rock. The process of fossilization preserves evidence of ancient life forms, and geologic interpretation of the enclosing sedimentary rock yields valuable information about the environments in which those ancient organisms lived. Paleontologists find more recently evolved organisms in the geologically younger layers of sedimentary rocks and more ancient life forms in the older layers of rocks. Original material (e.g., shell and/or bone) may be preserved as found, but chemical means may sometimes be used to alter or preserve it.

Radioactive dating provides another highly accurate method of confirming the age of rocks and fossils. Comparative anatomists study similarities and differences among organisms. Anatomists have been able to discover significant similarities in the skeletal architecture and musculature of all vertebrates from fish to humans. The most plausible explanation for this finding is that all vertebrates descended from a common ancestor.

> **3. d.** *Students know* how to construct a simple branching diagram to classify living groups of organisms by shared derived characteristics and how to expand the diagram to include fossil organisms.

Evolutionary relationships among living organisms and their ancestors can be displayed in a diagram that resembles a branching tree. Groups of similar living species belong to a genus, similar genera belong to a family, similar families belong to an order, similar orders belong to a class, and similar classes belong to a phylum. Working back in time from the shared derived characteristics of each living species contained in the diagram will show the evolutionary relationships leading to a common ancestor. The classification of organisms according to their characteristics is called *systematics*. It is based on a system developed in 1758 by the Swedish botanist and explorer Carolus Linnaeus.

3. e. *Students know* that extinction of a species occurs when the environment changes and the adaptive characteristics of a species are insufficient for its survival.

Extinction of a species occurs when the adaptive characteristics of the species are no longer sufficient to allow the species to survive under changing environmental conditions. Evidence from the fossil record indicates that most of the species that once lived on Earth are now extinct. Biological adaptations are produced through the evolutionary process. Random mutations in the DNA of different individuals (plants or animals) produce variations of particular traits in a population of organisms. These mutations result in some individuals acquiring characteristics that give them and their offspring an advantage in surviving and reproducing in their present environments or in a different environment. The offspring of individuals in which these advantageous characteristics are not present may decline in numbers and eventually become extinct, or they may continue to coexist with the offspring of individuals that have the mutational advantage. Natural selection will then lead to the existence of populations better able to survive and reproduce under any one particular environmental condition. However, when particular adaptive characteristics of a species are no longer sufficient for the survival of that species under changing environmental conditions (such as increased competition for resources, newly introduced predators, loss of habitat), that species may become extinct. There are many different environmental causes of the extinction of species.

STANDARD SET 4. Earth and Life History

The process of natural selection is strongly linked to the environment. Students will learn in this standard set that the environment has changed over time. The geologic record provides evidence of both the environments of the past and the plants and animals that inhabited them. The focus in this standard set is on using the geologic evidence to better understand life on Earth, past and present.

This standard set presents two great ideas from the geologic sciences to make clear the relationship between life and geology: (1) the concept of uniformitarianism; and (2) the principle of superposition. *Uniformitarianism* refers to the use of features, phenomena, and processes that are observable today to interpret the ancient geologic record. The idea is that small, slow changes can yield large cumulative results over long periods of time. Standard 4.c states a simplified version of the principle of *superposition* when it indicates that the oldest rock layers are generally found at the bottom of a sequence of rock layers. The principle of superposition is the basis for establishing relative time sequences (i.e., determining what is older and what is younger). Geologic records indicate that both local and global catastrophic events have occurred, including asteroid/comet impacts, that have significantly affected life on Earth. Both the evidence and the impact on life should be addressed in this standard set.

> **4. Evidence from rocks allows us to understand the evolution of life on Earth.** As a basis for understanding this concept:
>
> a. *Students know* Earth processes today are similar to those that occurred in the past and slow geologic processes have large cumulative effects over long periods of time.

This standard approaches two different but related ideas in the geologic sciences. The first (uniformitarianism) uses the present as the key to the past. For example, ripples preserved in ancient sedimentary rock are identical to ripples made by running water in mud and sand today. This idea is only one example of how geologists use the present to interpret features and processes in the geologic past. The second idea (superposition) states that the vastness of geologic time allows even very slow processes, if they continue long enough, to produce enormous effects. Perhaps the most important example of this idea is the dramatic change in the arrangement of the continents (continental drift) caused by the slow movement of lithospheric plates (approximately 5 centimeters per year) during the course of many millions of years. One piece of evidence for plate tectonics, including Pangaea, is the fossil record. The coherence of species in the fossil record is seen when geologic history is properly understood.

> **4. b.** *Students know* the history of life on Earth has been disrupted by major catastrophic events, such as major volcanic eruptions or the impacts of asteroids.

The subject of major catastrophic events is important because such events, although rare in the history of Earth, have had a significant effect on the shaping of Earth's surface and on the evolutionary development of life. Most of the time geologic processes proceed almost imperceptibly, only to be interrupted periodically by the impact of a large meteor or by a major volcanic eruption. The immediate effect of both types of catastrophic events is much the same: injection of large amounts of fine-grained particulate matter into the atmosphere, an event that may have immediate regional or even global consequences for the climate by causing both short- and long-term changes in habitats.

> **4. c.** *Students know* that the rock cycle includes the formation of new sediment and rocks and that rocks are often found in layers, with the oldest generally on the bottom.

Whenever rocks are uplifted and exposed to the atmosphere, they are subject to processes that can break them down. Purely physical processes, such as abrasion and freezing/thawing cycles, break rocks into smaller pieces. At the same time reactions with constituents of the atmosphere, principally acidic rain and oxygen, may cause chemical changes in the minerals that constitute the rocks and result in the formation of new types of minerals. The net result is called *sediment*. It consists of rock and mineral fragments, various dissolved ions, and whatever biological debris happens to be lying around. The sediment is removed by erosion from the sites where it formed and is transported by water, wind, or ice to other sites; there the sediment is deposited and eventually lithified to form new sedimentary rock. The biological portion of accumulated sediment may be fossilized and preserved, providing a partial record of existing life in the source area of the sediment.

Superposition, the fossil record, and related principles, such as crosscutting and inclusions, together form the basis for dating the relative ages of rocks. Students should realize that relative dating establishes only the order of events, not quantitative estimates of when those events actually occurred.

> **4. d.** *Students know* that evidence from geologic layers and radioactive dating indicates Earth is approximately 4.6 billion years old and that life on this planet has existed for more than 3 billion years.

Relative age-dating (see Standard 4.c) provides information about the relative sequence of events in the history of Earth. Absolute dating (putting a numerical estimate on the age of a particular rock sample) requires the use of a reliable "clock" in the form of the radioactive decay of certain naturally occurring elements. Those elements are disaggregated into the various minerals at the time those minerals are formed, generally during the crystallization of igneous rocks. Thus the newly formed minerals in the igneous rock contain only the original radioactive form of the element (parent) and none of the products of radioactive decay (daughter products), which are different from the parent. The rate of transformation by radioactive decay from parent to daughter elements can be measured experimentally. This rate is usually expressed as a *half-life*, which is defined as the amount of time it takes to change one-half of the atoms of the parent element to daughter products.

Earth's surface is always being reworked because of plate tectonics and erosion; therefore, very little of the planet's original material is available for dating. However, moon rocks and meteorites, thought to be the same age as Earth, can also be dated. All the available evidence points to Earth and the solar system being approximately 4.6 billion years old. The earliest rocks containing evidence of life are slightly more than 3 billion years old.

> **4. e.** *Students know* fossils provide evidence of how life and environmental conditions have changed.

Fossils provide evidence of the environments and types of life that existed in the past. As an ancient environment changed, so did the organisms it supported. Thus environmental changes are reflected by the classes of organisms that evolved during the period of environmental change. Uniformitarianism is the foundation on which these interpretations are based. For example, ancient animals exhibiting approximately the same shell shape and thickness as that of the modern clam probably lived in the same environment as clams do today. By examining fossil evidence and noting changes in life types over time, geologists can reconstruct the environmental changes that accompanied (perhaps caused) the changes in life types.

> **4. f.** *Students know* how movements of Earth's continental and oceanic plates through time, with associated changes in climate and geographic connections, have affected the past and present distribution of organisms.

Darwin's work on finches in the Galapagos Island demonstrated clearly the effect of isolation on the distribution of organisms. Geographic separation of individuals in a species prevented the populations from interbreeding. This separation may have led to the accumulation of genetic changes in the two populations, changes that eventually defined them as different species. Plate tectonic movements of lithospheric plates and the uplift of mountain ranges divided (albeit slowly) populations of plant and animal species and isolated the divisions from one another. This principle was illustrated in the fossil record of dinosaur species. Some dinosaurs, as well as other species that were restricted to specific continents after geologic separation, were uniformly distributed prior to continental separation.

> **4. g.** *Students know* how to explain significant developments and extinctions of plant and animal life on the geologic time scale.

Many changes that life has undergone during the history of Earth have been gradual, occurring as organisms adapt to slowly changing environments, evolve into new species, or become extinct. This principle is a fundamental tenet of uniformitarianism. But even uniformitarianism is not consistently true. For example, very early Earth on which the first life appeared was considerably different from the planet of today. Little oxygen was in the atmosphere, and no ozone layer was in the stratosphere to protect against harmful solar radiation. The earliest life was therefore anaerobic and had to be protected from solar radiation. Evidence for this single-celled life can be found in rocks that are slightly more than 3 billion years old.

Photosynthetic cyanobacteria, once referred to as blue-green algae, were an early addition to the prehistoric ecosystem. These early organisms are seen in the fossil record and were very successful, so much so that they still exist worldwide and are essentially unchanged in form after billions of years.

The slow change in Earth's life has been punctuated by sudden events—catastrophic ones—when viewed on the vast geologic time scale. One such remarkable event occurred about 600 million years ago. It is known as the Cambrian Explosion because of the sudden appearance of many different kinds of life, including many new multicellular animals that, for the first time, had preservable hard parts, such as shells and exoskeletons.

At various times life on Earth has also suffered from catastrophic mass extinctions in which the vast majority of species quickly died out. The greatest such event happened about 250 million years ago toward the end of the Paleozoic Era. It is known as the Permian extinction, and as much as 90 percent of marine species may have died out. Another famous mass extinction occurred at the end of the Mesozoic Era and is known as the Cretaceous-Tertiary (K-T) extinction. At the time all species of dinosaurs died out, as did about half of all the plant and animal groups. Evidence is mounting to indicate that this catastrophic event was caused by the impact of an asteroid.

STANDARD SET 5. Structure and Function in Living Systems

Students were first introduced in grade one to the complementary nature of structure and function when they studied the different shapes of animal teeth and inferred the kinds of food those animals eat. Students in grade three studied the external physical characteristics of organisms and considered their functions as a matter of adaptation. Students in grade seven will deepen their understanding of internal structures, a topic that was introduced in grade five.

Anatomists and physiologists consider at different levels the internal structures of living organisms. Mammals have discrete organs, many of which work together as systems. For example, the adrenal and pituitary glands are parts of the endocrine system, and the kidneys and bladder are parts of the excretory system. Flowering plants have tissues, such as xylem and phloem, that are part of a vascular system. Organs themselves may have specific tissues; for example, the white and gray matter of the brain can serve multiple functions. The pancreas produces both digestive enzymes and blood hormones.

Students in grade seven learn about the musculoskeletal system, the basic functions of the reproductive organs of humans, and the structures that help to sustain a developing fetus. Students also study the intricate structures of the eye and ear, which have well-understood functions in sight and hearing. Although many topics are covered in this section, they are all grouped in the fields of anatomy and physiology.

> **5. The anatomy and physiology of plants and animals illustrate the complementary nature of structure and function. As a basis for understanding this concept:**
>
> a. *Students know* plants and animals have levels of organization for structure and function, including cells, tissues, organs, organ systems, and the whole organism.

Protists, such as amoebae, consist of only one cell. All the functions necessary to sustain the life of these organisms must be carried out within that one cell. Multicellular organisms, such as plants and animals, tend to have cellular specialization (differentiation), which means individual cells or tissues may take on specific functions within the organism. For example, the musculoskeletal system of animals comprises individual muscle groups (e.g., biceps) that are bundles of muscle fibers, which are themselves groups of muscle cells, working together to make possible movements of the organism. Within individual muscle cells are organelles, such as the mitochondria, that help provide the energy for muscle contraction.

> **5. b.** *Students know* organ systems function because of the contributions of individual organs, tissues, and cells. The failure of any part can affect the entire system.

Students learned in grade five how blood circulates through the body and how oxygen, O_2, and carbon dioxide, CO_2, are exchanged in the lungs and tissues. The pulmonary–circulatory system functions as a whole because of the functions of its individual components. A person may die from a heart attack (from failure of the heart), suffocation or pneumonia (from insufficient gas exchange in the lungs), shock (from loss of blood volume), or a stroke (sometimes caused by an insufficient gas exchange with brain tissues due to the blockage of blood vessels).

> **5. c.** *Students know* how bones and muscles work together to provide a structural framework for movement.

The skeletal system in animals provides support and protection. Muscles are attached to bones by tendons and work in coordination with the bones and the nervous system to cause movement through coordinated contraction and relaxation of different muscle groups. For example, a muscle in the arm called the *biceps* causes bending of the arm at the elbow so that the angle between the bones (humerus and ulna) decreases. The *triceps* on the back of the arm causes bending so that the same angle increases. This flexion and extension of the arm is a good example of muscle groups that are coordinated. Even in a lifting motion in which one of those two muscle groups is ostensibly the prime mover of the bone (e.g., "curling" a weight with the biceps), the opposing muscle group is involved in producing a smooth, controlled motion of the arm and protecting the joint from strong contraction.

> **5. d.** *Students know* how the reproductive organs of the human female and male generate eggs and sperm and how sexual activity may lead to fertilization and pregnancy.

In males the testes in the external scrotum are the reproductive structures that produce sperm. Immature sperm cells in the walls of the seminiferous tubules of each testis mature into flagellated cells that are transported and stored in the epididymis. During sexual arousal millions of sperm may be transported to the urethra and ejaculated through the penis. Some sperm may exit through the penis before ejaculation (i.e., without the man's knowledge), and sexual activity that does not result in ejaculation may nonetheless lead to the release of sperm, fertilization, and pregnancy.

In females the ovaries are the reproductive structures that produce and store eggs, also called *oocytes* (pronounced "oh-oh-sights"). An egg develops within an ovarian structure called a *follicle*. A mature follicle can rupture through the wall of the ovary, releasing the egg during the process of ovulation. The egg is then transported by one of the Fallopian tubes to the uterus. If the female, at or around this time, engages in sexual activity that results in sperm being deposited in or near the vagina, a sperm cell can travel through the vagina to the uterus or Fallopian tubes and fertilize the egg. A fertilized egg may implant in the uterus and develop, meaning that the female is pregnant and may deliver a baby approximately nine months later. If the fertilized egg fails to implant and begin development, or if the egg is not fertilized, it will be sloughed off along with several layers of cells lining the uterus and leave the female's vagina during menstruation. One of the first signs of pregnancy is that a woman's regular monthly menstrual cycle stops.

> **5. e.** *Students know* the function of the umbilicus and placenta during pregnancy.

The placenta is an organ that develops from fetal tissue in the uterus during pregnancy. It is responsible for providing oxygen to the developing fetus. The umbilical cord (which enters the body at the *umbilicus,* or navel) is a cord containing arteries and veins that connect the fetus to the placenta. Although the blood of the mother and of her fetus do not mix together, oxygen and nutrients pass from the mother's blood to the fetus. Wastes, such as carbon dioxide from the fetus, are removed. The placenta helps to nourish and protect the fetus; however, most drugs and alcohol can easily pass from the mother's blood into the blood of the fetus, as can many infectious viruses, such as the human immunodeficiency virus (the source of AIDS).

> **5. f.** *Students know* the structures and processes by which flowering plants generate pollen, ovules, seeds, and fruit.

Flowering plants, or *angiosperms,* reproduce sexually by generating gametes in the form of sperms and ova. The reproductive structure of the angiosperms is the flower, which may contain male or female parts or both. Stamens are the male reproductive structures within the flower. Each stamen is composed of an *anther,* the structure that produces pollen granules, and a *filament,* the long thin stalk that connects the anther to the base of the flower (receptacle). The *pistil* is the female reproductive structure located in the center of the flower. The pistil consists of the *stigma,* which receives the pollen grains, and the *style,* a long thin stalk that acts as a guide for the pollen tube. The pollen tube, in turn, provides a migration path for the sperm of the pollen grain down to the ovary at the base of the pistil. The ovary contains one or more ovules, inside of which develop the ova. After fertilization the ovule develops into a seed with the developing embryo inside surrounded by a food source (the endosperm) for the plant embryo. The surrounding ovary may then enlarge and mature into a fruit that can contain one or more seeds.

5. g. *Students know* how to relate the structures of the eye and ear to their functions.

The eye works much like a camera. The eye is equipped with a lens that brings an image into focus on a sheet of light-sensitive cells called the *retina,* which is equivalent in a camera to a sheet of film or a video chip. The amount of light entering the eye is controlled by the iris, which is an adjustable circular aperture. In bright lighting the iris contracts and the pupil (the open area that appears black) becomes smaller in diameter to admit less light. In dim lighting the iris relaxes and the pupil becomes larger to admit more light. The lens of the eye refracts (or bends) the light, much as a magnifying glass does, and focuses an image on the retina. The lens is flexible, and its shape changes when focusing on nearby or distant objects. The retina contains cells that are sensitive to bright colors (cone cells) and others that are sensitive to dim lighting (rod cells). The cells in the retina generate an electrical signal that travels to the brain, which can interpret the visual pattern. Investigative activities with lenses may be practiced both in this standard set and in Standard Set 6, "Physical Principles in Living Systems," which describes the optics of sight.

The external ear (i.e., the part that can be seen) helps to collect sound waves and direct them to the middle ear. Many mammals (e.g., cats and many breeds of dogs) can redirect their external ears to detect faint sounds and determine the direction from which a sound is coming. The middle ear consists of a vibrating eardrum, or *tympanic membrane,* and three small bones (the *malleus,* or hammer; *incus,* or anvil; and *stapes,* or stirrup) that form a series of levers connecting the eardrum to the inner ear. Two small muscles control the tension on the eardrum and middle ear bones to reduce or increase the loudness of sound being transmitted. The inner ear, or labyrinth, contains the sensory cells that turn the waves of sound or pressure into electrical signals that are sent to the brain.

Students may explore the structure of the mammalian eye by performing a dissection. They should be able to identify and explain the function of the different parts of the eye. Students may learn the structure and function of the human ear by building a model from simple materials. Students should be able to identify the different parts of the ear and explain how those parts work together to transmit sensory information through sound waves. The sensory cells lining the cochlea are stimulated by the sound waves, causing nerve impulses to be transmitted through the auditory nerve to the brain.

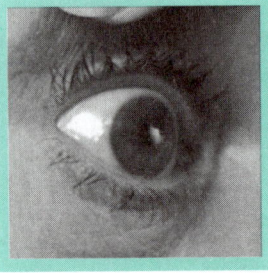

STANDARD SET 6. Physical Principles in Living Systems

The study of optics and levers, or pressure, is usually reserved for physical science classes. However, these topics are introduced for the first and only time in the seventh grade, so students should learn the principles behind them. Suggestions are made to relate the study of these topics to the eye, muscles, bones, tendons, and heart.

The human eye contains receptors that detect incoming visible light emitted by a luminous object or reflected from an illuminated object. Until the early 1900s physicists believed that the properties of light could be completely understood by viewing light as a wave of electromagnetic energy that was supported by an elusive medium—the so-called *ether*—that was imagined to pervade even a vacuum. The nature of light still seems mysterious to most people because light manifests the properties of both a wave and a particle. In most experiences geometric optics, which treats light as rays traveling in straight lines, adequately accounts for reflection and refraction, mirrors, and lenses. Before starting these topics, students should be able to measure angles, do ratio and proportion problems, and use gram mass weights and metersticks.

Students in grade seven can and should learn how levers confer a mechanical advantage. Given a lever, students should be able to identify the fulcrum and four important quantities: effort distance, effort force, resistance distance, and resistance force. If three of those quantities are known, students should be able to calculate the fourth quantity. Students can make simple levers and hinges (and other simple machines, if time permits) to show how levers can be used to increase forces at the expense of distances or distances at the expense of forces. Metersticks, weight holders, hooked weights, and pivoted supports are commercially available for students to make a straightforward investigation of the operation of levers. A key element of this standard set is to relate the physical principles to the function of muscle and bone in the body. Pressure, a subject that was introduced to students in the context of atmospheric pressure in earth science, is now discussed in the context of blood pressure and heart function.

> **6. Physical principles underlie biological structures and functions. As a basis for understanding this concept:**
>
> **a.** *Students know* visible light is a small band within a very broad electromagnetic spectrum.

Visible light is a part of a continuum known as the electromagnetic spectrum that extends on both sides of the visible region. This continuum includes the very long wavelengths, such as those of AM and FM radio and TV; the slightly shorter wavelengths, such as radar, microwave, and infrared radiation; and visible light that has wavelengths just less than one-millionth of a meter long. The wavelengths of electromagnetic radiation that the human eye can see vary from about 800 nanometers (0.0000008 m, or red light) to 400 nanometers (0.0000004 m, or blue/violet light). The colors of the visible spectrum are traditionally described as red, orange, yellow, green, blue, indigo, and violet but are actually a continuous spectrum.

> **6. b.** *Students know* that for an object to be seen, light emitted by or scattered from it must be detected by the eye.

This standard deals with the physical principles of the interaction of light with matter. After the initial interaction light rays from an object must pass from the object to the eye. The interactions with those parts of the eye that focus the light, creating an image on the retina, and transfer the light into electrical impulses, which are interpreted by the brain, all depend on the information in the light that enters the eye. This information arises from the initial interaction of the light with the object or the nature of the light emitted by the light source(s) or both. The color and brightness of the light that is emitted or reflected from an object depend on the color, brightness, and angle of incidence from the source illuminating the object. The object then absorbs, reflects, or refracts the illuminating light and imparts a color and brightness. That color is attributed to the object, but color really depends on the source of light and the way the object interacts with it.

This process scatters light in all directions. The eye detects only the light that enters it. This light first encounters the front, rounded, transparent surface of the eye (the cornea), where most of the focusing occurs. Next, it enters the interior of the eyeball through the pupil and passes through the lens, which acts to further focus the light to accommodate both near and far objects. The focused light then falls on the receptors (the rod cells and the cone cells) in the retina, is converted into electrical impulses, and is transferred by the optic nerve to the visual cortex of the brain.

> **6. c.** *Students know* light travels in straight lines if the medium it travels through does not change.

In a vacuum or in a uniformly transparent material, light travels in straight lines. At the interface between two media or between a vacuum and a medium, light rays will bend if they enter at an angle other than perpendicular to the interface. The light-bending properties of objects should be explored. However, transparent materials, such as air, may have differing densities and cause light to bend as it passes through the material. For example, the air heated by a campfire can cause objects to appear to shimmer because the path of the light is not a straight line. The variations in the density of the atmosphere are what cause the stars to twinkle. When light travels from one transparent medium (such as air) into another transparent medium with different optical properties (such as water), the path of the light may bend (or be refracted) depending on the angle of the ray of light in relation to the surface between the two media.

A pencil placed in a glass half full of water will appear bent. By analyzing the path of the light from various points on the pencil to the eye of the observer, students will be able to confirm that the path of the light did change direction as it passed from one medium into another.

> **6. d.** *Students know* how simple lenses are used in a magnifying glass, the eye, a camera, a telescope, and a microscope.

Combinations of lenses are used in telescopes and microscopes to magnify objects. The cornea of the eye plays the major role of a lens in transforming the rays of light diverging from an object into rays of light converging to a focus on the retina. To provide instruction in this standard, teachers may use magnifiers. Simple magnifiers of plastic (or glass) are inexpensive and easily obtained. A magnifier is a converging optic because it can convert rays of light diverging from an object to rays of light converging to form an image. Magnifiers are characterized by their focal lengths, which may be found by lifting a lens up from a table until the sharpest image of a ceiling light is formed. The distance from the magnifier to the image on the tabletop is the focal length. If the magnifier is held at a distance shorter than the focal distance above a printed page, the print is seen magnified because the lens creates an enlarged, virtual image instead of a real image. If the magnifier is held at a distance greater than the focal length above the page, what is seen depends on where the observer's eye is located. The light leaving the lens is now converging so that if the eye intercepts the converging rays, no sharp image will be seen. If the eye is located far enough above the page, the rays from the lens converge to form a real image and pass through it. The eye is now intercepting diverging rays and sees the print upside down.

> **6. e.** *Students know* that white light is a mixture of many wavelengths (colors) and that retinal cells react differently to different wavelengths.

White (visible) light may be dispersed into a spectrum of colors: from red at the longest wavelength to violet at the shortest wavelength. A glass or plastic prism disperses white light into the colors of the spectrum because the angle of refraction is different for each of the different wavelengths (colors). A diffraction grating of closely spaced grooves can also be used to separate white light into various colors because different wavelengths (i.e., different colors) interfere constructively after reflection at different angles. Teachers should present both these effects to show the nature of white light.

The human perception of color is due to specialized color light receptor cells in the retina of the eye. These specialized cells (called cone cells) make color vision possible. Full-color printing is achieved by the use of just four ink colors (usually magenta, yellow, and cyan along with a very dark purple or black). The four colors are printed in combinations of dot patterns too small to perceive (resolve) with the human eye. Color images in magazines are commonly produced in this way.

> **6. f.** *Students know* light can be reflected, refracted, transmitted, and absorbed by matter.

The interaction of light with matter may be classified as reflection, refraction, transmission, or absorption. Light transmitted through air and transparent, uniform materials continues to travel in a straight line. However, when rays of light encounter a surface between two materials or two media, such as air and water or air and glass, the light may be reflected or refracted at the surface. The angle at which the light is reflected or refracted from its original path follows principles that depend on the optical properties of the materials, such as the angle of incidence being equal to the angle of reflection. The principles of refraction are what make it possible for lenses to focus and magnify images.

Light travels (is transmitted) through a transparent medium by a process of absorption and reemission of the light energy by the atoms of the medium. Opaque and translucent objects absorb and scatter light from their original direction much more strongly than do transparent objects. Optically denser materials, such as glass, cause light to travel more slowly than do less optically dense materials, such as water and especially air. Light travels through air just slightly more slowly than through a vacuum. Rays of light may be observed to change direction, or refract (a consequence of light changing speed), in going from one medium to another. However, if light enters a new medium perpendicular to its surface, the light continues in a straight line so that refraction is not observed (even though the light is traveling at a different speed in the second medium). Impurities or imperfections in transparent materials or media cause some of the light to be scattered out of a beam. Smoke, fog, and clouds decrease visibility because they scatter light.

> **6. g.** *Students know* the angle of reflection of a light beam is equal to the angle of incidence.

When a light beam encounters a shiny reflecting surface, the angle of reflection is the same as the angle of incidence. The angle is usually measured in relation to the surface normal.

> **6. h.** *Students know* how to compare joints in the body (wrist, shoulder, thigh) with structures used in machines and simple devices (hinge, ball-and-socket, and sliding joints).

Archimedes is credited with first understanding that a rigid rod (a lever) able to rotate about a fixed pivot point (a fulcrum) can be used to turn a small force into a large force. Joints in the body act as pivot points for bones acting as levers, and muscles provide the force. There are three classes of levers, which are defined by the relative positions of the applied force causing the action, the placement of the fulcrum, and the resistant object being moved. A lever provides one of two principal advantages: It can amplify the force being applied so that a small force applied over a long distance can create a large force over a short distance. This principle is useful to know in lifting heavy objects. The alternative is typical of levers in the human body: A large force applied over a short distance in a short time can be amplified into long, rapid motions, such as in running or in swinging a baseball bat.

> **6. i.** *Students know* how levers confer mechanical advantage and how the application of this principle applies to the musculoskeletal system.

A lever can be used to take advantage of force or speed (or motion). A bone is the lever; a joint is the pivot point (or fulcrum); muscles supply the force; and connective tissues transfer the force to locations that usually give an individual the leverage to increase his or her speed of motion of foot, arm, or hand. Students can make simple levers and hinges (and other simple machines, if time permits) to show how levers may be used to increase force at the expense of distance or distance at the expense of force. Metersticks, weight holders, hooked weights, and pivoted supports are commercially available for students to make straightforward investigations of the operation of levers. These or other hands-on laboratory activities using first-, second-, and third-class levers in simple equipment will make the "law of the lever" more real than will solving a set of mathematical proportion problems or merely identifying the parts of a lever from drawings or pictures.

> **6. j.** *Students know* that contractions of the heart generate blood pressure and that heart valves prevent backflow of blood in the circulatory system.

The heart is a pump in which blood enters a chamber through a blood vessel; a valve closes off the blood vessel to prevent the blood from flowing in the wrong direction; and the heart muscle contracts. This action "squeezes" the blood and increases the pressure to force the blood into another blood vessel. Pressure is defined as force per unit area and is measured in various units, such as millimeters of mercury (mmHg). Students may learn more about the physiology of the heart by reading science texts and studying models.

STANDARD SET 7. Investigation and Experimentation

The essential skills and knowledge of observation, communication, and experimental design are extended in grade seven. Including scale model building in the curriculum helps students to visualize complex structures. Collecting information from a variety of resources is an important part of scientific inquiry and experimental design. Many types of print and electronic resources are available in the school library to support teaching and learning science. The skills needed to search out and recognize accurate and useful resources are complex and generally require significant knowledge of the topic.

Chapter 4
The Science Content Standards for Grades Six Through Eight

Grade Seven
Focus on Life Sciences

7. **Scientific progress is made by asking meaningful questions and conducting careful investigations. As a basis for understanding this concept and addressing the content in the other three strands, students should develop their own questions and perform investigations. Students will:**

   a. Select and use appropriate tools and technology (including calculators, computers, balances, spring scales, microscopes, and binoculars) to perform tests, collect data, and display data.

   b. Use a variety of print and electronic resources (including the World Wide Web) to collect information and evidence as part of a research project.

   c. Communicate the logical connection among hypotheses, science concepts, tests conducted, data collected, and conclusions drawn from the scientific evidence.

   d. Construct scale models, maps, and appropriately labeled diagrams to communicate scientific knowledge (e.g., motion of Earth's plates and cell structure).

   e. Communicate the steps and results from an investigation in written reports and oral presentations.

Grade Eight Focus on Physical Sciences

Students in grade eight study topics in physical sciences, such as motion, forces, and the structure of matter, by using a quantitative, mathematically based approach similar to the procedures they will use in high school. Earth, the solar system, chemical reactions, the chemistry of biological processes, the periodic table, and density and buoyancy are additional topics that will be treated with increased mathematical rigor, again in anticipation of high school courses. Students should begin to grasp four concepts that help to unify physical sciences: force and energy; the laws of conservation; atoms, molecules, and the atomic theory; and kinetic theory. Those concepts serve as important organizers that will be required as students continue to learn science. Although much of the science called for in the standards is considered "classical" physics and chemistry, it should provide a powerful basis for understanding modern science and serve students as well as adults.

Mastery of the eighth-grade physical sciences content will greatly enhance the ability of students to succeed in high school science classes. Modern molecular biology and earth sciences, as well as chemistry and physics, require that students have a good understanding of the basics of physical sciences.

STANDARD SET 1. Motion

Aristotle wrote that a force is required to keep a body moving. Everyday experience seems to confirm this misconception. For two thousand years Aristotle's description of motion was accepted without question. Then an experiment by Galileo resulted in the discovery of friction. Galileo's experimental approach to investigating Nature helped to establish modern science and led to the invention of calculus and Newton's laws of motion. Four centuries after Galileo the knowledge of motion enables scientists to predict and control the paths of distant spacecraft with great accuracy.

There are many types of motion: straight line, circular, back and forth, free-fall, projectile, orbital, and so on. This standard set concerns itself with the motion of a body traveling either at a constant speed or with a varying speed that is represented by an average value.

> **1. The velocity of an object is the rate of change of its position. As a basis for understanding this concept:**
>
> **a.** *Students know* position is defined in relation to some choice of a standard reference point and a set of reference directions.

The position of a person or object must be described in relation to a standard reference point. For example, the position of a bicycle may be in front of the flagpole or behind the flagpole. The flagpole is the reference point, and *in front of* and *behind* are the reference directions. A reference point is usually called the *origin,* and position can be expressed as a distance from the reference point together with a plus (+) or minus (−) sign that may stand for *in front of* and *behind, away from* and *toward, right* and *left,* or one of any other pair of convenient, opposing directions from the reference point.

The idea of measuring positions, distances, and directions in relation to a standard reference point may be introduced by using metersticks (or rulers). The students are directed to call the 50 cm mark (or some other convenient mark) the reference point. A position of −10 cm would be 10 cm to the left of the standard reference point; a position of +5 cm would be 5 cm to the right of the standard reference point. The teacher may call out various positive and negative position values, and the students should point to that location on the ruler. In particular, students can experience the fact that although moving in a positive direction (to the right) when going from −10 cm to −6 cm, they still end up pointing to a spot that is to the left of the origin. Students in grade eight should be able to track the motion of objects in a two-dimensional (*x, y*) coordinate system. For example, both *x* and *y* may represent distances along the coordinate axes, or the value of *y* might represent the distance traveled and *x* might represent elapsed time.

> **1. b.** *Students know* that average speed is the total distance traveled divided by the total time elapsed and that the speed of an object along the path traveled can vary.

Speed is how fast something is moving in relation to some reference point without regard to the direction. It is calculated by dividing the distance traveled by the elapsed time. In the next standard students should learn to use the International System of Units (a modernized version of the metric system) to measure distance in meters (m) and time intervals in seconds (s). Thus a car traveling 120 kilometers in two hours is traveling at a speed of 60 km/hr. (In everyday units speed is measured in miles per hour. In the school laboratory it may be more convenient to use centimeters instead of meters for measuring distances and seconds for measuring time; therefore, speed would be expressed in centimeters per second [cm/s].) The speed of a spacecraft may be measured by how long it takes to orbit Earth and the length of that orbit. Sometimes the speed of an object remains constant while it is being observed, but usually the speed of a vehicle changes during a trip. Students should be taught to recognize that the average speed of a vehicle is calculated by dividing the total distance traveled by the length of time to complete the entire trip. With several stops a trip of 100 miles from town A to town B may take four hours. The average speed is 100 miles ÷ 4 hours = 25 miles per hour (mph) even though at times the car may have had a speedometer reading of 55 mph.

Students can measure the entire distance that a toy vehicle or ball travels across the floor or tabletop after it is released from the top of an inclined ramp (the standard reference point). They can also measure the time elapsed during the trip. The average speed can then be calculated by dividing the distance traveled (from the standard reference point) by the elapsed time. More than one student may be assigned to measure the times and distances so that duplicate data sets are created. The teacher may explore with the students why the data sets are not exactly the same and help them evaluate the accuracy and reproducibility of the experiment. The object's speed may be observed to change during the trip: it travels faster down the ramp because of gravity and slows down as it travels across the floor or tabletop because of friction. What is being calculated by $v = d/t$ (where v is the average speed, d is the total distance traveled, and t is the elapsed time) is the average speed for the entire trip as though the object were to travel at a constant speed. Students may change one of the conditions, such as the height of the ramp, to see how that affects the average speed. Or students do not have to wait for the object to stop; they may measure the elapsed time for the object to roll from the top of the ramp to any point along the path, before the object stops, to obtain the average speed between the measurement points.

> **1. c.** *Students know* how to solve problems involving distance, time, and average speed.

Problems related to this standard may be solved by using the traditional mathematics formula: $d = rt$. The d represents the total distance traveled, r stands for rate (or speed) and represents either the constant speed (if the speed is constant) or average speed (if it varies), and t represents the time taken for the trip. Given any two of these quantities, students can calculate the third quantity: $d = rt$, $t = d/r$, $r = d/t$. Students may be given information involving d, r, or t for different segments of a real or hypothetical trip and asked to use the formula $d = rt$ to solve for the missing information. To avoid confusion later, teachers may introduce the symbol v for speed instead of r once students are familiar with this type of problem. (When the vector nature of *velocity* needs to be introduced, the v will be written in boldfaced type, **v**, as will other vector quantities in the framework.)

> **1. d.** *Students know* the velocity of an object must be described by specifying both the direction and the speed of the object.

The word *velocity* has a special meaning in science. An air traffic controller needs to know both the speed and the direction of an aircraft (as well as its position), not just the speed. Measurable quantities that require both the magnitude (sometimes the term *size* is used) and direction are called *vector quantities*. Displacement, velocity, acceleration, and force are all vector quantities and will be introduced in grade eight by using only one dimension or specified pathway. An arrow pointing in the direction of motion usually represents the velocity of an object. The length of the arrow is proportional to how fast the object is going (the speed). Students demonstrate mastery of this standard by knowing, without prompting, that they must specify both speed and direction when asked to describe an object's velocity.

> **1. e.** *Students know* changes in velocity may be due to changes in speed, direction, or both.

Since velocity is a vector quantity, the velocity of an object is determined by both the speed and direction in which the object is traveling. Changing the speed of an object changes its velocity; changing the direction in which an object is traveling also changes the velocity. A change in either speed or direction (or both) will, by definition, change the velocity. (Although the term is not included in this standard set, the rate at which velocity changes with time is called *acceleration.* When a car speeds up or slows down, it undergoes acceleration. When a car rounds a curve maintaining the same speed, it also undergoes acceleration because it changes direction.)

The important idea is that a change in the speed of the object, the direction of the moving object, or both is a change in velocity. Students may easily understand that a change in the speed of an object causes a change in the velocity; it may be less obvious to students that a change in the direction of an object, with no change in the speed, also changes the velocity of the object. Students need to recognize that spinning, curving soccer balls, baseballs, or Ping-Pong balls may maintain a nearly constant speed through the air but change velocity because they change direction. Of course, an object may undergo a change in velocity in which both the direction and the speed change; for example, when a driver applies the brakes while going around a curve.

In the next standard set, students will learn that changes in velocity are always related to one or more forces acting on the object. Students learn to find and identify forces and to determine the direction of each force's action. Being able to recognize velocity changes of magnitude and direction is key to observing and characterizing forces.

> **1. f.** *Students know* how to interpret graphs of position versus time and graphs of speed versus time for motion in a single direction.

Students are required to apply the graphing skills they learned in lower grades to the plotting and interpretation of graphs of distance, location, and position (d) versus time (t) and of speed (v) versus time (t) for motion in a single direction. A major conceptual difference from the graphing skills learned in mathematics is that the two axes will no longer be number lines with no units. What must be explicitly addressed in dealing with motion graphs is the plotting of locations in distance units (e.g., meters, centimeters, miles) on the vertical axis and plotting of time in time units (seconds, minutes, hours) on the horizontal axis.

In plotting position versus time, students should learn that the vertical axis represents distances away from an origin either in the positive (+) or negative (−) direction. The horizontal axis represents time. Every data point lying on the horizontal axis is "at the origin" because its distance value is zero. Given a graph of position versus time, students should be able to generate a table and calculate average speeds for any time interval ($v = d/t$). If the graph of position versus time is a straight line, the speed is constant; students should be able to find the slope and know that the slope of the line is numerically equal to the value of the speed in units corresponding to the labels of the axes.

Students should know that a graph of speed versus time consisting of a horizontal line represents an object traveling at a constant speed, and they should be able to use $d = rt$ to calculate the distance (d) traveled during a time interval (t). Students should know that a graph of speed versus time that is not a horizontal line indicates the speed is changing.

STANDARD SET 2. Forces

The concept of force is central to the study of all natural phenomena that involve some kind of interaction between two or more objects regardless of whether visible motion occurs. For example, architects and civil engineers want their structures to stand firm against the forces of gravity, wind, and earthquakes. On the other hand, automotive engineers need to know how best to accelerate a car, brake it to a safe stop, and smoothly change its direction. Students need to know that balanced forces keep an object from changing its velocity and that changes in the velocities of objects are caused by unbalanced forces.

There are only four known fundamental forces: gravitational forces, electromagnetic forces, and two nuclear forces known as the strong and the weak forces. Gravitational force is the attraction all objects with mass have for one another. The common experience of gravity on Earth is only one example; the other forces of pushing and pulling are elastic forces caused by electromagnetic interactions between atoms and molecules being pushed together or pulled apart. The large, repulsive electrical forces between the positively charged protons in the nucleus of an atom are balanced against the stronger, attractive nuclear forces that hold the atom together.

Students learned in grade two that the way to change how something is moving is to give it a push or a pull (e.g., apply a force). In grade four the study of magnets, compasses, and static electricity gave students experience with electromagnetic forces. In grade seven students learned about motion and forces, which involved comparing bones, muscles, and joints in the body to machines.

2. Unbalanced forces cause changes in velocity. As a basis for understanding this concept:

a. *Students know* a force has both direction and magnitude.

Forces are pushes or pulls and, like velocity, are vector quantities described by the magnitude and the direction of a force. As noted in Standard 1.d, the direction and strength of a force may be indicated graphically by using an arrow. The length of the arrow is proportional to the strength of the force, and the arrow points in the direction of the force's application.

The simplest case to consider is that of forces acting along one line, such as to the left or to the right. These colinear forces act either in the positive direction and are represented as positive quantities or in the negative direction and are represented as negative quantities.

A worthwhile activity is to have the students pull objects across level surfaces to measure the forces of friction. Different surfaces, because of varying roughness or different types of material, will exert different forces of friction on an object being dragged across them. If an object is pulled at a constant speed across a level surface, the force applied is just equal and opposite to the force of friction. If the force applied is greater than the force of friction, the object will slide easily. If the force applied is less than the force of friction, the object will drag. If the force applied is zero, the object will slow down and stop more quickly under the influence of the force of friction alone. Students can obtain data by using a spring scale to measure the force and compare different objects on different surfaces.

> **2. b.** *Students know* when an object is subject to two or more forces at once, the result is the cumulative effect of all the forces.

Forces acting on an object along the same line at the same time are calculated by using algebra. For example, a force of 5 newtons acting in the positive direction (+5 N) and a force of 7 newtons acting in the negative direction (–7 N) will result in an unbalanced force of 2 newtons acting in the negative direction (–2 N). A force of one newton is close to the weight of half a stick of butter or of a small apple. (In high school physics, students will learn that forces acting at different angles on an object can be broken down into components along the *x* axis, *y* axis, and *z* axis and that these components can also be calculated algebraically.)

> **2. c.** *Students know* when the forces on an object are balanced, the motion of the object does not change.

When several forces act simultaneously on an object, they may amount to zero, meaning there is no net force on the object and the motion of the object does not change. For example, a force of 10 newtons acting to the right (+10 N) and a second force of 10 newtons acting to the left (–10 N) amount to zero, meaning there will be no change in the velocity of the object. Sometimes an object acted on by balanced forces is at rest and remains at rest. In a tug of war in which opposing sides are pulling a rope with equal force, the rope does not move.

Sometimes a moving object is acted on by balanced forces and continues to move at the same velocity. For example, pushing a book straight across a table at a constant velocity requires force. The book does not speed up, slow down, or change direction; therefore, one must conclude a frictional force is pushing back on the book. Many people have the misconception that a force is necessary for an object to maintain a constant velocity; they overlook the opposing force of friction. Identifying and analyzing the forces acting on a sliding object by observing its velocity can help students develop their observation and analysis of frictional forces.

If the motion (or velocity) of an object is not changing, one may conclude that all the forces must be balanced. There are two equal and opposing vertical forces (weight down and table up) acting on the book as well as two equal and opposing horizontal forces (sliding push and friction): a total of four forces.

> **2. d.** *Students know* how to identify separately the two or more forces that are acting on a single static object, including gravity, elastic forces due to tension or compression in matter, and friction.

The force of gravity pulls objects toward the center of the earth. This force of gravity is commonly called the *weight* of the object. If an object is dropped, the force of gravity alone causes the velocity of the object to increase rapidly in the down direction. But when a single object is at rest, such as a book on a table, the table must be supplying a balancing upward force (an elastic force of compression caused by the compacting of the molecules of the table). When an object, such as a yo-yo, is observed hanging motionless from a string, the string must be supplying a balancing upward force—an elastic force of tension as its molecules are stretched apart. A student may push gently on a book to move it horizontally across the table, but the book does not move. The horizontal push cannot be the only acting force. A second force pushes back to keep the book at rest. This opposing force is the friction between the molecules in the surface of the book and the surface of the table.

Resting a book on a meterstick spanning the gap between two student desks usually causes the meterstick to sag, showing that the meterstick flexes until the upward force from its elastic distortion is sufficient to support the book. Resting a book on a soft, dry sponge or spring might also show how elastic forces support the book against the downward pull of gravity.

> **2. e.** *Students know* that when the forces on an object are unbalanced, the object will change its velocity (that is, it will speed up, slow down, or change direction).

When an unbalanced force acts on an object initially at rest, the object moves in the direction of the applied force. If an object is already in motion, for example, traveling to the right, and an unbalanced force acts to the right, the object will speed up. An object traveling to the right acted on by an unbalanced force to the left will slow down; if the unbalanced force continues to act, the object may slow to a stop and even begin to move faster in the opposite direction. If an unbalanced force acts in a direction perpendicular to the direction the object is moving, the force will deflect the object from its path, changing its direction but not its speed along the curved path. Any force that acts in such a direction (for example, the force of the road on the tires of a car) is called a *centripetal force.* This force is directed to the center of the orbit. Finally, an unbalanced force acting at an angle to the path may affect both the speed and the direction of the object.

Students should be able to predict changes in velocity if forces are shown to be acting on an object and be able to identify that an unbalanced force is acting on an object if they observe a change in its velocity. Students may not be able to explain fully the cause of the unbalanced forces acting on the baseball pitcher's curve ball or on the path of a spinning soccer ball, but they can state that there is a force acting perpendicular to the path of the ball.

> **2. f.** *Students know* the greater the mass of an object, the more force is needed to achieve the same rate of change in motion.

When the forces acting on an object are unbalanced, the velocity of the object must change by increasing speed, decreasing speed, or altering direction. This principle also means that if an object is observed to speed up, slow down, or change direction, an unbalanced force must be acting on it. The rate of change of velocity is called *acceleration.* At the high school level, students will learn to solve problems by using Newton's second law of motion, which states that the acceleration of an object is directly proportional to the force applied to the object and inversely proportional to its mass. For now students should learn to recognize acceleration (or deceleration) and should be able to state the direction and relative magnitude of the force that is the cause of the acceleration.

When an unbalanced force acts on an object, the velocity of the object can change slowly or rapidly. How fast the velocity of the object changes, that is, the rate of change in velocity with time (called acceleration), depends on two things: the size of the unbalanced force acting on the object and the mass of the object. The larger the unbalanced force, the faster the velocity of the object changes, but the greater the mass of the object, the slower the velocity changes. Quantitatively, the acceleration of an object may be predicted by dividing the net force acting on the object by the mass of the object.

Often high school students learn to solve problems involving force without clearly relating the physical circumstances to the word problem presented. It is important to teach students in grade eight to identify mass, velocity, acceleration, and forces and to analyze how those factors relate to one another in the physical system being studied. The ability to make qualitative predictions about what will happen next in these situations is the key to successful problem solving that all scientists use before starting a calculation. Once the correct qualitative prediction is envisioned, a numerical solution is more likely to be correct. For example, students might be told that an opposing force is applied to an object being pushed along the ground. Given all the numbers needed to calculate the object's final velocity, the students should be able to predict correctly whether the object could slow down, come to a stop, or even start moving backward before they solve the problem numerically.

> **2. g.** *Students know* the role of gravity in forming and maintaining the shapes of planets, stars, and the solar system.

Gravity, an attractive force between masses, is responsible for forming the Sun, the planets, and the moons in the solar system into their spherical shapes and for holding the system together. It is also responsible for internal pressures in the Sun,

Earth and other planets, and the atmosphere. Newton asked himself whether the force that causes objects to fall to Earth could extend to the Moon. Newton knew that the Moon should travel in a straight line (getting farther and farther from Earth) unless a force was acting on it to change its direction into a circular path.

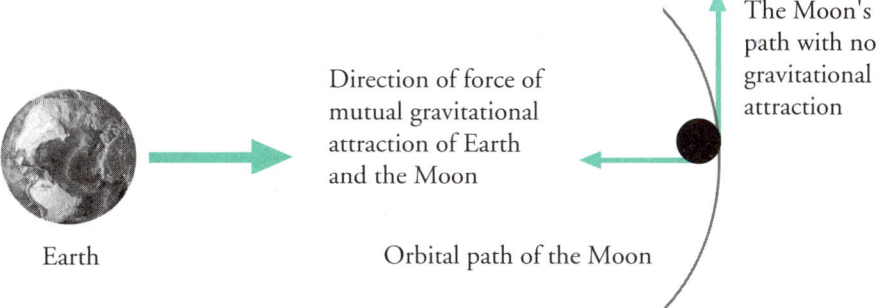

Fig.1. Effect of Gravity on the Moon's Path

He worked out the mathematics that convinced him that the force between all massive objects is directly proportional to the product of their masses and inversely proportional to the square of the distance between their centers. This relationship was then extended to explain the motion of Earth and other planets about the Sun.

Initially, the universe consisted of light elements, such as hydrogen, helium, and lithium, distributed in space. The attraction of every particle of matter for every other particle of matter caused the stars to form, making possible the "stuff" of the universe. As gravity is the fundamental force responsible for the formation and motion of stars and of the clusters of stars called galaxies, it controls the size and shape of the universe.

STANDARD SET 3. Structure of Matter

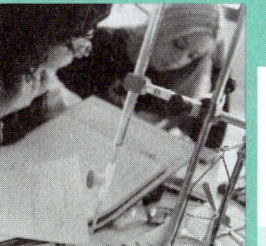

There is no disagreement about the importance of understanding the structure of matter. Richard Feynman, a famous Nobel prize-winning physicist, has said:

> If, in some cataclysm, all scientific knowledge were to be destroyed and only one sentence passed on to the next generation of creatures, what statement would contain the most information in the fewest words? I believe it is the atomic hypothesis (or atomic fact, or whatever you wish to call it) that all things are made of atoms—little particles that move around in perpetual motion attracting each other when they are a little distance apart, but repelling upon being squeezed into one another.[7]

Teachers should assess students' knowledge prior to instruction of this topic, as the atomic theory of matter may be very challenging to them. Students are expected to recall terms and definitions from earlier introductions to the concepts of atoms, molecules, and elements. Instruction should provide empirical evidence for the atomic theory, which will be useful for understanding science and crucial to the study of chemistry.

When students learn about the structure of matter, teachers should emphasize that the historical evidence for atoms was based largely on indirect measurements and inferences far removed from direct experience. Recently, instruments have been built that produce images of individual atoms, confirming what was inferred earlier as a result of overwhelming evidence from many scientific experiments. Most scientists come to know the atomic theory is true by repeatedly using the concepts and principles presented in the theory to explain observed properties and predict changes in matter.

> **3. Each of the more than 100 elements of matter has distinct properties and a distinct atomic structure. All forms of matter are composed of one or more of the elements. As a basis for understanding this concept:**
>
> **a.** *Students know* the structure of the atom and know it is composed of protons, neutrons, and electrons.

Shortly after British physicist Ernest Rutherford inferred the existence of atomic nuclei, the general idea emerged that atoms are mostly empty space with a tiny, massive nucleus at the center containing positively charged protons and neutral neutrons. This nucleus is surrounded by tiny, negatively charged electrons, each with about 1/2,000 the mass of a proton or neutron. Danish physicist Niels Bohr developed a model of the hydrogen atom to explain its visible spectrum. At the high school level, the chemistry standards require students to know the historical importance of this model. Bohr's model succeeded in predicting the spectrum of light emitted by hydrogen atoms and is therefore the acknowledged starting point for understanding atomic structure. However, Bohr's "solar" model of the atom, diagrammed in most textbooks as showing electrons in circular orbits about the nucleus, is oversimplified. Rather than try to describe how the electrons in an atom are moving, teachers are better advised to help students develop a model of the atom in which each electron has definite energy. Students should know that the energy of each electron in an atom keeps it in motion around the positive nucleus to which it is attracted. The structure of multielectron atoms is understood in terms of electrons filling energy levels that define *orbitals.*

> **3. b.** *Students know* that compounds are formed by combining two or more different elements and that compounds have properties that are different from their constituent elements.

The word *combining* implies bonding. Understanding the concepts of ionic and covalent bonding helps explain why some elements combine to form compounds and some do not. Atoms of different elements combine to form compounds; a compound may, and usually does, have chemical characteristics and physical properties that are different from those of its constituent elements. Examples and generalizations may be drawn from ionic compounds formed of metals and nonmetals and covalently bonded, organic compounds formed from carbon and other elements.

Students often learn to manipulate chemical equations without having a picture in their minds of physical reality at the atomic level. The ability to create such a picture is a useful skill that helps students keep track of all the atoms in the process. For example, the reaction of methane and oxygen to form carbon dioxide and water can be visualized by using models or drawing pictures of the atoms and molecules in the reactants. These molecules can then be rearranged into new products. (Make sure that all the atoms in the starting reactants are accounted for in the new products.) Instruction in this standard will help students understand that compounds are collections of two or more different kinds of atoms that are bonded together. Knowing exactly how the atoms are organized to form a molecule is not essential.

> **3. c.** *Students know* atoms and molecules form solids by building up repeating patterns, such as the crystal structure of NaCl or long-chain polymers.

Crystals of table salt, the compound NaCl, have a regular, cubic structure in which sodium (Na^+) ions alternate with chlorine (Cl^-) ions in three-dimensional array with the atoms at the corner of cubes forming the lattice. In organic polymers, the carbon, hydrogen, sometimes oxygen, and nitrogen atoms combine to form long, repetitive, stringlike molecules.

Inexpensive models of molecules may be made by using colored gumdrops (held together by toothpicks) to represent molecules. Students identify the atoms that constitute the molecules by using a color-coded key relating the color of the gumdrop to an atom of an element. They learn that the shape of a molecule is important to its chemical and physical properties. At the high school level, students will be introduced to the idea that shape is determined mainly by the electron configuration that provides the most energy-stable system.

Students can also grow crystals from a solution and should understand that this process leads to the building up of atoms into a lattice. Students may begin the process by dissolving an excess of sodium chloride, sugar, or Epsom salts in water. Then they hang a string in the water and store the container in a place where it will be undisturbed while the water evaporates. Crystals will form on the string. Putting a small (seed) crystal tied to a piece of thread in the solution will accelerate the growth process. Books and kits (including chemicals, glassware, and instructions) on crystal growing are available commercially. Students can watch crystals grow on slides under a microscope. Some crystals display vivid colors when viewed between crossed sheets of polarizing material.

> **3. d.** *Students know* the states of matter (solid, liquid, gas) depend on molecular motion.

All atoms, and subsequently all molecules, are in constant motion. For any given substance the relative freedom of motion of its atoms or molecules increases from solids to liquids to gases. When a thermometer is inserted into a substance and the temperature is measured, the average atomic or molecular energy of motion is being measured. The state of matter of a given substance therefore depends on the balance between the internal forces that would restrain the motion of the atoms or molecules and the random motions that are in opposition to those restraints.

The change in phases is evidence of various degrees of atomic and molecular motion. The conditions of temperature and pressure under which most materials change from solid to liquid or liquid to vapor (gas) or gas to plasma have been measured. Those properties are difficult to predict but are highly reproducible for different samples of the same material and can be used to identify substances. Some substances will go from solid to gas directly at one atmosphere pressure. Dry ice, which is frozen carbon dioxide, is an example. Chemistry handbooks contain the melting points (or freezing points) and boiling points (or condensation temperatures) of most materials usually under one atmosphere pressure. If the pressure is not one atmosphere, those temperatures change. Some substances have more than one stable solid phase at room temperature. Graphite, with its soft black texture and its hard, clear crystalline diamond atomic structure, represents the two solid phases of elemental carbon.

Water is another example of a substance that undergoes a change in atomic and molecular motion under extreme conditions of temperature and pressure. At one atmosphere pressure, ice forms when water is cooled below zero degrees Celsius (or 32 degrees Fahrenheit). Above the freezing point the average molecular energy of motion of the water molecules is just enough to overcome the attractive forces between the molecules. The water molecules thereby avoid being locked in place and remain liquid. At and below the freezing point, the water molecules become the solid, crystalline material called ice. When liquid water is heated to temperatures of 100 degrees Celsius, molecular motion increases until large groups of water molecules overcome the attractive forces between the molecules. At this point those energetic molecules form bubbles of steam, which are bubbles of gas made not of air but of water. The process in which bubbles of water vapor escape from liquid water is called *boiling*. Continued heating will change the liquid water entirely into vapor instead of raising the temperature of the water above 100 degrees Celsius.

> **3. e.** *Students know* that in solids the atoms are closely locked in position and can only vibrate; in liquids the atoms and molecules are more loosely connected and can collide with and move past one another; and in gases the atoms and molecules are free to move independently, colliding frequently.

The atoms or molecules of a solid form a pattern that minimizes the structural energy of the solid consistent with the way in which the atoms or molecules attract at long distances but repel at short distances. The atoms or molecules vibrate about their equilibrium positions in this pattern. When raised above the melting temperature, the atoms or molecules acquire enough energy to slide past one another so that the material, now a liquid, can flow; the density of the liquid remains very close to that of the solid, demonstrating that in a solid or a liquid the atoms stay at about the same average distance.

If a single atom or molecule acquires enough energy, however, it can pull away from its neighbors and escape to become a molecule of a gas. Gas molecules move about freely and collide randomly with the walls of a container and with each other. The distance between molecules in a gas is much larger than that in a solid or a liquid, and this point may be emphasized when students study density.

> **3. f.** *Students know* how to use the periodic table to identify elements in simple compounds.

The periodic table of elements is arranged horizontally in order of increasing atomic number (number of protons) and vertically in columns of elements with similar chemical properties. Students should learn to use the periodic table as a quick reference for associating the name and symbol of an element in compounds and ions. They should be able to find the atomic number and atomic weight of the element listed on the table. The periodic table is both a tool and an organized arrangement of the elements that reveals the underlying atomic structure of the atoms. This standard focuses on the table as a tool.

Every field of science uses the periodic table, and various forms of it exist. Astrophysicists may have a table that includes elemental abundances in the solar system. Physicists and engineers may use tables that include boiling and melting points or thermal and electrical conductivity of the elements. Chemists have tables that show the electron structures of the element. Students should be encouraged to refer to the periodic table as they study the properties of matter and learn about the atomic model.

STANDARD SET 4. Earth in the Solar System (Earth Sciences)

Students in grade eight are ready to tackle the larger picture of galaxies and astronomical distances. They are ready to study stars compared with and contrasted to the Sun and to learn in greater detail about the planets and other objects in the solar system. High school studies of earth sciences will include the dimension of time along with three-dimensional space in the study of astronomy.

> **4.** **The structure and composition of the universe can be learned from studying stars and galaxies and their evolution. As a basis for understanding this concept:**
>
> **a.** *Students know* galaxies are clusters of billions of stars and may have different shapes.

Stars are not uniformly distributed throughout the universe but are clustered by the billions in galaxies. Some of the fuzzy points of light in the sky that were originally thought to be stars are now known to be distant galaxies. Galaxies themselves appear to form clusters that are separated by vast expanses of empty space. As galaxies are discovered they are classified by their differing sizes and shapes. The most common shapes are spiral, elliptical, and irregular. Beautiful, full-color photographs of astronomical objects are available on the Internet, in library books, and in popular and professional journals. It may also interest students to know that astronomers have inferred the existence of planets orbiting some stars.

> **4. b.** *Students know* that the Sun is one of many stars in the Milky Way galaxy and that stars may differ in size, temperature, and color.

The Sun is a star located on the rim of a typical spiral galaxy called the Milky Way and orbits the galactic center. In similar spiral galaxies this galactic center appears as a bulge of stars in the heart of the disk. The bright band of stars cutting across the night sky is the edge of the Milky Way as seen from the perspective of Earth, which lies within the disk of the galaxy. Stars vary greatly in size, temperature, and color. For the most part those variations are related to the stars' life cycles. Light from the Sun and other stars indicates that the Sun is a fairly typical star. It has a mass of about 2×10^{30} kg and an energy output, or luminosity, of about 4×10^{26} joules/sec. The surface temperature of the Sun is approximately 5,500 degrees Celsius, and the radius of the Sun is about 700 million meters. The surface temperature determines the yellow color of the light shining from the Sun. Red stars have cooler surface temperatures, and blue stars have hotter surface temperatures. To connect the surface temperature to the color of the Sun or of other stars, teachers should obtain a "black-body" temperature spectrum chart, which is typically found in high school and college textbooks.

> **4. c.** *Students know* how to use astronomical units and light years as measures of distance between the Sun, stars, and Earth.

Distances between astronomical objects are enormous. Measurement units such as centimeters, meters, and kilometers used in the laboratory or on field trips are not useful for expressing those distances. Consequently, astronomers use other units to describe large distances. The astronomical unit (AU) is defined to be equal to the average distance from Earth to the Sun: 1 AU = 1.496×10^{11} meters. Distances between planets of the solar system are usually expressed in AU. For distances between stars and galaxies, even that large unit of length is not sufficient. Interstellar and intergalactic distances are expressed in terms of how far light travels in one year, the light year (ly): 1 ly = 9.462×10^{15} meters, or approximately 6 trillion miles. The most distant objects observed in the universe are estimated to be 10 to 15 billion light years from the solar system. Teachers need to help students become familiar with AUs by expressing the distance from the Sun to the planets in AUs instead of meters or miles. A good way to become familiar with the relative distances of the planets from the Sun is to lay out the solar system to scale on a length of cash register tape.

> **4. d.** *Students know* that stars are the source of light for all bright objects in outer space and that the Moon and planets shine by reflected sunlight, not by their own light.

The energy from the Sun and other stars, seen as visible light, is caused by nuclear fusion reactions that occur deep inside the stars' cores. By carefully analyzing the spectrum of light from stars, scientists know that most stars are composed primarily of hydrogen, a smaller amount of helium, and much smaller amounts of all the other chemical elements. Most stars are born from the gravitational compression and heating of hydrogen gas. A fusion reaction results when hydrogen nuclei combine to form helium nuclei. This event releases energy and establishes a balance between the inward pull of gravity and the outward pressure of the fusion reaction products.

Ancient peoples observed that some objects in the night sky wandered about while other objects maintained fixed positions in relation to one another (i.e., the constellations). Those "wanderers" are the planets. Through careful observations of the planets' movements, scientists found that planets travel in nearly circular (slightly elliptical) orbits about the Sun.

Planets (and the Moon) do not generate the light that makes them visible, a fact that is demonstrated during eclipses of the Moon or by observation of the phases of the Moon and planets when a portion is shaded from the direct light of the Sun.

Various types of exploratory missions have yielded much information about the reflectivity, structure, and composition of the Moon and the planets. Those missions have included spacecraft flying by and orbiting those bodies, the soft landing of spacecraft fitted with instruments, and, of course, the visits of astronauts to the Moon.

> **4. e.** *Students know* the appearance, general composition, relative position and size, and motion of objects in the solar system, including planets, planetary satellites, comets, and asteroids.

Nine planets* are currently known in the solar system: Mercury, Venus, Earth, Mars, Jupiter, Saturn, Uranus, Neptune, and Pluto. They vary greatly in size and appearance. For example, the mass of Earth is 6×10^{24} kg and the radius is 6.4×10^{6} m. Jupiter has more than 300 times the mass of Earth, and the radius is ten times larger. The planets also drastically vary in their distance from the Sun, period of revolution about the Sun, period of rotation about their own axis, tilt of their axis, composition, and appearance. The inner planets (Mercury, Venus, Earth, and Mars) tend to be relatively small and are composed primarily of rock. The outer planets (Jupiter, Saturn, Uranus, and Neptune) are generally much larger and are composed primarily of gas. Pluto is composed primarily of rock and is the smallest planet in the solar system.

All the planets are much smaller than the Sun. All objects are attracted toward one another gravitationally, and the strength of the gravitational force between

---

* Under resolutions passed by the International Astronomical Union on August 24, 2006, there are eight planets. Pluto no longer meets the definition of a "planet" but is now classified under a new distinct class of objects called "dwarf planets."

them depends on their masses and the distance that separates them from one another and from the Sun. Before Newton formulated his laws of motion and the law of universal gravitational attraction, German astronomer Johannes Kepler deduced from astronomical observations three laws (Kepler's laws) that describe the motions of the planets.

Planets have smaller objects orbiting them called *satellites* or *moons*. Earth has one moon that completes an orbit once every 28 days (approximately). Mercury and Venus have no moons, but Jupiter and Saturn have many moons. Very small objects composed mostly of rock (asteroids) or the ice from condensed gases (comets) or both also orbit the Sun. The orbits of many asteroids are relatively circular and lie between the orbital paths of Mars and Jupiter (the asteroid belt). Some asteroids and all comets have highly elliptical orbits, causing them to range great distances from very close to the Sun to well beyond the orbit of Pluto.

Teachers should look for field trip opportunities for students to observe the night sky from an astronomical observatory or with the aid of a local astronomical society. A visit to a planetarium would be another way of observing the sky. If feasible, teachers should have students observe the motion of Jupiter's inner moons as well as the phases of Venus. Using resources in the library-media center, students can research related topics of interest.

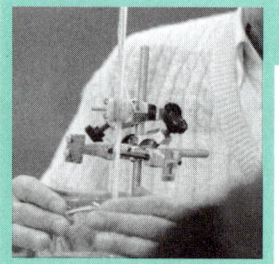

STANDARD SET 5. Reactions

When substances react, the atoms involved in the reactants are rearranged, forming other products. Students have learned that the physical and chemical properties of the newly formed substances (products) are different from the physical and chemical properties of the original substances (reactants). Students in grade eight will learn that it is the underlying arrangement of the atoms in the reactants and products and the energy needed or released during the rearrangement process that explain chemical reactions. Understanding chemical reactions is essential because they constitute, directly or indirectly, a large portion of the discipline of chemistry.

Students need to be able to distinguish a chemical change from a physical change. In a physical change one or more physical properties of the material are altered, but the chemical composition (i.e., the arrangement of the atoms in molecules) remains the same. In a chemical change the atoms are rearranged to form new substances with different chemical and physical properties. Students must be familiar with the periodic table and the names and symbols of the chemical elements.

In grade one students are prepared for the idea of chemical reactions when they learn that the properties of substances can change when they are mixed, cooled, or heated. In grade three they learn that when two or more substances are combined, a new substance may be formed with properties that are different from those of the original materials. In grade five they learn that during chemical reactions, the atoms in the reactants rearrange to form products with different properties.

The study on reactions begun in grade eight will support future studies about conservation of matter and stoichiometry as well as work on acids, bases, and solutions. Students will go beyond studying reactions and their reactant/product relationships to work with the rates of reaction and chemical equilibrium. Students should be able to envision a chemical equation at the atomic and molecular levels. They should "see" the number of reactant atoms and molecules in the equation coming together and by some process rearranging into the correct number of atoms and molecules that form the products. This important conceptual skill helps students to keep track of all the atoms.

> **5. Chemical reactions are processes in which atoms are rearranged into different combinations of molecules. As a basis for understanding this concept:**
> **a.** *Students know* reactant atoms and molecules interact to form products with different chemical properties.

This standard focuses on changes that occur when atoms and molecules as reactants form product compounds with different chemical properties. Teachers may have students perform simple reactions or demonstrate the reactions for students. All students should be able to learn the more important chemical reactions and the elements involved in them, especially if common compounds such as vinegar (acetic acid), baking soda (sodium bicarbonate), table salt (sodium chloride), carbonated water, and nutritional minerals and foods are used in activities or demonstrations. An example might involve adding calcium chloride and baking soda to water. Such reactions demonstrate clearly the differences in properties between reactants (solids and liquid) and products (solid, liquid, and gas).

> **5. b.** *Students know* the idea of atoms explains the conservation of matter: In chemical reactions the number of atoms stays the same no matter how they are arranged, so their total mass stays the same.

The conservation of matter is a classical concept, reinforcing the idea that atoms are the fundamental building blocks of matter. Atoms do not appear or disappear in traditional chemical reactions in which the constituent atoms and/or polyatomic ions are simply rearranged into new and different compounds. Conservation of atoms is fundamental to the idea of balancing chemical equations. The total number of atoms of each element in the reactants must equal the total number of atoms of each element in the products. The total number of atoms, hence the total mass, stays the same before and after the reaction.

There are several ways to teach and assess students' understanding of the concept of conservation of mass in chemical reactions. Weighing reactants before and products after a reaction shows that mass is neither gained nor lost. However, experimental errors are possible; the most common one is not sufficiently drying the products before weighing. One simple demonstration of the concept that atoms (or matter) are conserved in chemical reactions in which mass might appear to be lost is to determine the combined mass of a small, sealed container filled one-third with water, the screw-on cap, and one-quarter of an effervescent tablet. After the piece of tablet is dropped in the water, the container is immediately sealed. When the fizzing has stopped, the combined mass of the sealed container and the tablet should remain the same. After the seal is broken, much of the carbon dioxide gas formed by the reaction escapes, and the mass of the container and its contents decrease.

Students should also be taught to balance simple chemical equations. This step reinforces the idea that atoms do not appear or disappear in chemical reactions and, therefore, that matter is conserved.

5. c. *Students know* chemical reactions usually liberate heat or absorb heat.

In chemical reactions the atoms in the reactants rearrange to form products, and there is usually a net change in energy. Breaking bonds between atoms requires energy; making a bond releases energy. If the total making and breaking of all bonds for a particular chemical reaction results in a net release of energy, the reaction is said to be *exothermic*. The energy is typically released as heat into nearby matter. If the total making and breaking of bonds results in a net absorption of energy, the reaction is called *endothermic*. The energy is typically absorbed as heat from nearby matter, which therefore cools. A convenient way to demonstrate that chemical reactions release or absorb heat is the application of the hot packs or cold packs used for athletic injuries. The change in temperature produced by those packs may be the result of a chemical reaction, or it may be caused by a "heat of solution" and not by a chemical reaction. For example, dissolving is considered a physical and not a chemical change because the compound may be recovered, unchanged chemically, by evaporation.

5. d. *Students know* physical processes include freezing and boiling, in which a material changes form with no chemical reaction.

When heated, many solid materials undergo a reversible change of state into a liquid (melting). Under the standard condition of one atmosphere of pressure, the temperature at which such a solid material melts is the same as the temperature at which the liquid material freezes; this temperature, called the *melting point,* is characteristic of the material. Many liquid materials when heated also undergo a reversible change of state into a gas. Under one atmosphere of pressure, such a liquid material may boil; the temperature at which this occurs is also characteristic of the material and is called the *boiling point.* Such reversible changes—back and forth from solid to liquid or from liquid to gas—are called physical changes because no chemical change (a permanent reordering of the atoms into new molecules) occurs. Similarly, the dissolving of one substance into another, such as a solid or gas into a liquid, is often reversible (by evaporating the liquid to leave the solid or heating the liquid to drive out the gas) and is also called a physical rather than a chemical change. Physical changes can usually be undone to recover the original materials unchanged. Activities such as mixing iron filings with sand demonstrate a physical change. In this case a magnet can recover the iron filings from the mixture.

5. e. *Students know* how to determine whether a solution is acidic, basic, or neutral.

Indicators that change color are routinely used to determine whether a solution is acidic, basic, or neutral. A pH scale indicates with numbers the concentration of hydrogen ions in a solution and characterizes a solution as acidic (lower than 7), basic (higher than 7), or neutral (near 7). There are electrodes and electronic instruments that can measure directly the pH of a solution. Some acids and bases are defined other than by their hydrogen ion concentration, but they will be addressed in high school chemistry. Teachers may give students the opportunity to test solutions, including foods such as fruits and vegetables, with pH paper, litmus paper, indicator solutions, or pH meters to determine whether a solution or food is acidic, basic, or neutral. Students should be familiar with the pH scale to know what a given pH value indicates.

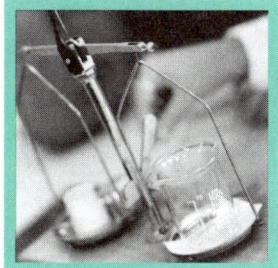

STANDARD SET 6. Chemistry of Living Systems (Life Sciences)

Because all living organisms are made up of atoms, chemical reactions take place continually in plants and animals, including humans. The uniqueness of organic chemistry stems from *chain polymers.* Life could not exist without the ability of some chemicals to join together, repetitively, to form large, complex molecules. Concepts learned in this standard set are critical for understanding fully the chemistry of the cells of organisms, genetics, ecology, and physiology that will be taught in the high school biology/life sciences standard sets.

6. Principles of chemistry underlie the functioning of biological systems. As a basis for understanding this concept:

a. *Students know* that carbon, because of its ability to combine in many ways with itself and other elements, has a central role in the chemistry of living organisms.

Carbon is unique among the elements because it can bond to itself and to many other elements. This attribute makes possible many different kinds of large, carbon-based molecules. Typically, carbon will make four separate covalent bonds (to other carbon atoms), but double and triple bonds are also possible. The variety of bonds allows carbon-based molecules to have a wide range of shapes and chemical properties. Key shapes include tetrahedral (e.g., methane and carbon tetrachloride), planar (e.g., formaldehyde and ethylene), and linear (e.g., acetylene and carbon dioxide). Students can research the nomenclature, composition, and structure of organic molecules by using textbooks and supplemental instructional materials. They can also construct models of carbon-based molecules by using commercial modeling kits or inexpensive alternatives (e.g., gumdrops and toothpicks).

> **6. b.** *Students know* that living organisms are made of molecules consisting largely of carbon, hydrogen, nitrogen, oxygen, phosphorus, and sulfur.

Living organisms are made up of a great variety of molecules consisting of many atoms (with carbon atoms playing the main roles), but the number of different elements involved is quite small. Carbon and only five other elements make up most of Earth's biomass. Those six elements, however, can combine in many different ways to make large, organic molecules and compounds. To demonstrate this idea, teachers may burn organic material, such as bone, leaves, wood, or a variety of candles. They may hold a cold glass or plate above the flame to condense droplets of water, one of the combustion products. They may also hold a heat-treated glass in the flames to collect carbon deposits in the form of soot. Students can discuss what elements were in the organic material. Teachers may draw students' attention to the black material that forms when meat is roasted or grilled or when toast is charred.

> **6. c.** *Students know* that living organisms have many different kinds of molecules, including small ones, such as water and salt, and very large ones, such as carbohydrates, fats, proteins, and DNA.

Living organisms require a variety of molecules; some molecules contain carbon and some do not. The molecules that make up organisms and control the biochemical reactions that take place within them are usually large molecules, such as DNA, proteins, carbohydrates, and fats. Organisms also require simple substances, such as water and salt, to support their functioning. Teachers may encourage students to research why plants and animals need simple molecules such as water. Other activities for teachers may include squeezing the water from celery or turnips to demonstrate the presence of water. Or they may ask students how they can demonstrate that water is in fruits and vegetables (e.g., dried fruit). Teachers may also ask students how they know that there is salt in their bodies. Most students know that their perspiration tastes salty.

STANDARD SET 7. Periodic Table

Students will need to know the chemical symbols of the common elements. It will be helpful for them to be familiar with other properties of materials, such as melting temperatures, boiling points, density, hardness, and thermal and electrical conductivity. By the time students begin the study of this standard set, they should be familiar with the periodic table and should know the names and chemical symbols of most of the common elements. In this standard set they must now look in greater detail at and learn the significance of atomic numbers and isotopes and how they relate to the classification of elements. Students need to go more deeply into the elemental properties that serve as the basis for the periodic arrangement. Meeting the standards in this set will serve as a strong foundation for the study of atomic and molecular structures and of the relationship between these structures and the arrangement of elements on the periodic table that will take place in high school chemistry.

A common form of the periodic table has 18 columns (groups of elements) in the main body. This form shows the periodicity, or repeating pattern, of chemical and some physical properties of the elements. What varies most in published periodic tables is the information provided in the box that represents each element. The most useful tables are those that show the physical properties of the most common form of the element in addition to the atomic number and the atomic weight. A table that color-codes metals and nonmetals is also useful.

Elements shown toward the top of the periodic table are lighter, and those toward the bottom are heavier. Elements shown to the left are generally metallic, and those toward the right are nonmetallic. The word *metallic* refers to the collective properties of common metals: luster, malleability, high electrical and thermal (heat) conductivity. Although the majority of elements in the periodic table are metals, a few are classified as semimetals and may be found bordering the transition between the metals and nonmetals. When atoms from the left side of the table combine with atoms from the right side, they tend to form ionic salts, which are brittle crystalline compounds with high melting temperatures.

At the high school level, students will learn that the arrangement of the elements in the columns of the periodic table reflects the electron structure of the atoms of each element. This pattern explains the similarity in the chemical properties of the elements in each column of the periodic table.

Students should be readily able to use the periodic table to find the atomic number of an element and should know that there is a pattern of increasing atomic numbers as the table is read from left to right and down one row at a time. The lanthanides and actinides are placed off the table to save space; however, if they were placed in the table they would still be read in the same manner—from left to right and then down. Students should also know that the atomic number is the number of protons in the nucleus.

> **7. The organization of the periodic table is based on the properties of the elements and reflects the structure of atoms. As a basis for understanding this concept:**
>
> **a.** *Students know* how to identify regions corresponding to metals, nonmetals, and inert gases.

The periodic table of elements is structured so that metals are shown on the left, with the most reactive metals on the far left. Nonmetals are located on the right, with the most reactive next to the "inert" gases on the far right. Despite the name, inert gases are not truly inert. Although no naturally occurring inert gas compounds are known, some have been synthesized in the laboratory. Therefore most scientists use the term *noble* gas instead of inert gas. Semimetals, found between the metals and nonmetals in the periodic table, are elements, such as silicon, that have some

Chapter 4
The Science Content Standards for Grades Six Through Eight

Grade Eight
Focus on Physical Sciences

properties of metals but also have properties that are typical of nonmetals. Although only a few elements fit this category, the unique electrical property of semimetal elements is that they are semiconductors, an essential property for computer chips. The rare earth elements can be used to produce very strong magnets.

Students should know that scientists have the right to name their discoveries and that some elements have been named after famous men and women scientists, such as curium, einsteinium, and seaborgium.

> **7. b.** *Students know* each element has a specific number of protons in the nucleus (the atomic number) and each isotope of the element has a different but specific number of neutrons in the nucleus.

A rigorous definition of the term *element* is based on the number of protons in the atom's nucleus (the atomic number). All atoms of a given element have the same number of protons in the nucleus. Atoms with different atomic numbers are atoms of different elements. Although the number of protons is fixed for a particular element, the same is not true for the number of neutrons in the nucleus. An element that has different numbers of neutrons in its atoms is called an *isotope* of the element. For example, all hydrogen atoms have one proton in the nucleus, but there are two additional isotopes of hydrogen with different numbers of neutrons. One is called deuterium (one proton and one neutron), and the other is called tritium (one proton and two neutrons). The common isotope of hydrogen has one proton and no neutrons in its nucleus.

Some isotopes are *radioactive,* meaning that the nucleus is unstable and can spontaneously emit particles or trap an electron to become the nucleus of a different element with a different atomic number. All the isotopes of some elements are radioactive, such as element 43, technetium, or element 86, radon. No stable samples of those elements exist. Element 92, uranium, is another example of an element in which no stable isotopes exist. However, uranium (atomic weight 238) is found in nature because it decays so slowly that it is still present in Earth's crust. The atomic number of each element represents the number of protons in the nucleus. Therefore, as the atomic number increases, the mass of the atoms of succeeding elements generally increases although exceptions exist because of the varying numbers of neutrons in some isotopes. Typically, however, the atoms of the elements in the periodic table increase from left to right, and those elements listed in the lower rows are more massive than those in the upper rows.

> **7. c.** *Students know* substances can be classified by their properties, including their melting temperature, density, hardness, and thermal and electrical conductivity.

The physical properties of substances reflect their chemical composition and atomic structure. The melting temperature or hardness of the common forms of the elements is related to the forces that hold the atoms and molecules together. One can compare the boiling points of carbon and nitrogen. Carbon is solid up to very high temperatures (3,600 degrees Celsius); nitrogen, the element next to it, is a gas until it is cooled to below negative 196 degrees Celsius. This dramatic difference between two adjacent elements on the periodic table shows there must be very different intermolecular forces acting as a result of a slight change in atomic structure.

Density is the mass per unit volume and is a function of both the masses of individual atoms and the closeness with which the atoms are packed.

Electrical conductivity and thermal conductivity are strongly dependent on how tightly electrons are held to individual atoms. Metals and nonmetals may be found in portions of the periodic table. Metal atoms combine in regular patterns in which some electrons are free to move from atom to atom, a condition that accounts for both high electrical and high thermal conductivity.

STANDARD SET 8. Density and Buoyancy

The central goal of this standard set is to be able to answer the simple question, Will an object sink or will it float? Students will learn that density is a physical property of a substance independent of how much of the substance is available, and they will be able to relate the property of density to the phenomenon of buoyancy. Archimedes, a Greek mathematician, is credited with first recognizing that different substances have different densities and that fluids exert a buoyant force on objects submerged in them. He came to this understanding while trying to determine what else was in a supposedly gold crown.

Archimedes came to a simple realization: water does not sink in water. That is, if one focuses on one drop of water in a container of water (and if one could keep the drop intact and distinct), the drop would not fall to the bottom of the container even though it has weight. The surrounding water must exert an upward buoyant force on the drop equal to the weight of the drop. The drop would fall if its weight were greater than the buoyant force supplied by the surrounding water. The drop would rise if it weighed less than the buoyant force. The surrounding water exerts an upward buoyant force on any volume within it equal to the weight of that volume of water. Understanding the nature of floating and sinking led Archimedes to realize that different substances have different densities—the key to determining whether the crown was gold or a fake.

Density is a property characteristic of the material itself and does not change whether the material is subdivided or the amount available is altered. Different substances have different densities, so knowing the density of a sample is useful in determining its composition. For example, the composition of Earth's interior was first inferred to be different from the composition of rocks in the lithosphere because the density of lithospheric rocks is different from the average density of Earth.

The density of solids and liquids, the two condensed states of matter, does not vary much with changes in pressure or temperature. However, small differences in density within a liquid or gas may be caused by local heating and result in convection currents. Because gases are so compressible, their densities may vary over a wide range of values. That is why tables of measured values of density are found only for solids and liquids.

Most fluids (gases and liquids) are very poor conductors of heat. Normally, fluids expand when their temperature increases because of the more rapid motion of the constituent molecules. If a fluid is heated locally, the thermal energy is not conducted rapidly to other parts of the fluid; the region that is hotter expands, becoming less dense than the cooler surrounding fluid. The buoyant force supplied by the cooler surrounding fluid on the hotter expanded region is greater than the weight of the hotter region. This force causes the hotter, less dense region of fluid to be pushed up, a phenomenon known as "Hot air rises."

A thorough understanding of density and buoyancy will be helpful in mastering the earth science standards in high school.

> **8. All objects experience a buoyant force when immersed in a fluid. As a basis for understanding this concept:**
>
> **a.** *Students know* density is mass per unit volume.

Density is a physical property of a substance independent of the quantity of the substance. That is, a cubic centimeter of a substance has the same density as a cubic kilometer. Density may be expressed in terms of any combination of measurements of mass and volume. The measurement units most commonly used in science are grams per cubic centimeter for solids and grams per milliliter for liquids.

> **8. b.** *Students know* how to calculate the density of substances (regular and irregular solids and liquids) from measurements of mass and volume.

Density is calculated by dividing the mass of some quantity of material by its volume. Mass may be determined by placing the material on a balance or scale and subtracting the mass of its container. The volume of a liquid may be measured easily by using graduated cylinders, and the volume of a regular solid may be measured by using a ruler and the appropriate geometry formula. It is not as simple, however, to measure the volume of an irregular solid. The volume of an irregularly shaped solid object may be determined by water displacement.

> **8. c.** *Students know* the buoyant force on an object in a fluid is an upward force equal to the weight of the fluid the object has displaced.

Whether an object will float depends on the magnitude of the buoyant force of the surrounding fluid (liquid or gas) compared with the weight of the object. The buoyant force is equal to the weight of the volume of fluid displaced by the object. The net force acting on a submerged body is the difference between the upward buoyant force of the surrounding fluid and the downward pull of gravity on the object (its weight). The same relationship applies to two separate fluids of differing densities. Therefore, if the volume of the fluid displaced by a submerged solid object weighs more than the object, the object will rise to the surface and float. If the values are the same, the object is said to be neutrally buoyant and will neither sink nor rise to the surface. If the volume of the surrounding fluid displaced by a solid object weighs less than the object, the object will sink.

The buoyant force can be demonstrated convincingly by placing a water-filled, sealed plastic sandwich bag in a container of water and noting that the sandwich bag filled with water does not sink even though gravity applies a downward force on the water-filled bag (its weight). Therefore, there must be an upward, buoyant force applied by the surrounding water. If the sandwich bag is filled with a liquid that weighs more than an equal volume of water, it will sink. If the liquid in the sandwich bag is less dense than water, it will float. Students can fill another sandwich bag with hot water to demonstrate that it floats in room-temperature water. They can fill a third sandwich bag with water slightly above the freezing point and repeat the experiment to show that cold water will sink in room-temperature water.

To demonstrate buoyant forces dramatically, the teacher may place a heavy object, such as a large rock, in a large container of water and ask students first to lift the object in the container without removing the object from the water. Then the teacher asks students to lift the object completely from the water. Students are usually startled by how much easier it is to lift the object while it is in a container of water than to lift that same object when it is on a dry surface. People can move heavy stones from one place to another when the stones are immersed in rivers and lakes, but they often cannot lift the stone from the water. A small beach ball pushed down into a large container of water produces the same effect in reverse for students who have never experienced the large buoyant force that water can exert on a volume that is mostly air.

> **8. d.** *Students know* how to predict whether an object will float or sink.

The most direct way to predict whether a substance or solid object will sink or float in a fluid is to compare the density of the substance or object with the density of the fluid, either by measurement or by looking up the values on a table of densities. If the object is less dense than the fluid, it will float. Materials with densities greater than that of a liquid can be made to float on the liquid (e.g., steel boats and concrete canoes floating on water) if they can be shaped to displace a volume of the liquid equal to their weight before they submerge completely.

The density of liquids may be determined by using a hydrometer, either one that is commercially available or one that is made from a pencil with a thumbtack in the eraser. The depth to which an object of uniform density will sink in a liquid is a relative measure of the density of the liquid. Simple hydrometers, based on this principle, can be used to compare the densities of a variety of liquids with the density of water. The length of the hydrometer submerged in an unknown liquid *(U)* compared with the length submerged in water *(W)* can be used to determine the density of an unknown liquid *(W/U)* in metric units of grams per cubic centimeter. How far a pencil hydrometer sinks in water may be marked as "1 gram per cubic centimeter." If the pencil sinks twice as far in another liquid, its density is 0.5 gram per cubic centimeter; if it sinks half as far, the density is 2 grams per cubic centimeter; and so on.

Air is also a fluid and exerts a buoyant force on objects submerged in it. Hot-air balloons rise because the upward buoyant force of the cooler surrounding air is greater than the weight of the hot, less dense air inside the balloon and the trappings of the balloon. Helium balloons rise because a volume of helium gas is much lighter than an equal volume of air at the same temperature and pressure.

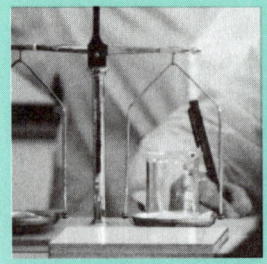

STANDARD SET 9. Investigation and Experimentation

Experiments can yield consistent, reproducible answers, but the answers may be incorrect or off the mark for many reasons. By the time students complete grade eight, they should have a foundation in experimental design and be able to apply logical thinking processes to evaluate experimental results and conclusions. Mathematical representation of data is the key to making quantitative scientific predictions. Graphs expressing linear relationships utilize proportional reasoning and algebra. Students should be taught to apply their knowledge of proportions and algebra to the reporting and analysis of data from experiments.

9. **Scientific progress is made by asking meaningful questions and conducting careful investigations. As a basis for understanding this concept and addressing the content in the other three strands, students should develop their own questions and perform investigations. Students will:**

   a. Plan and conduct a scientific investigation to test a hypothesis.

   b. Evaluate the accuracy and reproducibility of data.

   c. Distinguish between variable and controlled parameters in a test.

   d. Recognize the slope of the linear graph as the constant in the relationship $y = kx$ and apply this principle in interpreting graphs constructed from data.

   e. Construct appropriate graphs from data and develop quantitative statements about the relationships between variables.

   f. Apply simple mathematical relationships to determine a missing quantity in a mathematic expression, given the two remaining terms (including speed = distance/time, density = mass/volume, force = pressure × area, volume = area × height).

   g. Distinguish between linear and nonlinear relationships on a graph of data.

Notes

1. *Reading/Language Arts Framework for California Public Schools, Kindergarten Through Grade Twelve.* Sacramento: California Department of Education, 1999; *Mathematics Framework for California Public Schools, Kindergarten Through Grade Twelve* (Revised edition). Sacramento: California Department of Education, 2000.

2. *Mathematics Content Standards for California Public Schools, Kindergarten Through Grade Twelve.* Sacramento: California Department of Education, 1999.

3. *Science Safety Handbook for California Public Schools.* Sacramento: California Department of Education, 1999.

4. *Health Framework for California Public Schools, Kindergarten Through Grade Twelve.* Sacramento: California Department of Education, 1994.

5. Charles Darwin, *On the Origin of Species by Means of Natural Selection.* Reprinted from the 6th edition. New York: Macmillan, 1927.

6. Thomas R. Malthus, *An Essay on the Principle of Population.* 1798. Reprint. Amherst, N.Y.: Prometheus Books, 1998.

7. Richard P. Feynman, Robert B. Leighton, and Matthew Sands, Vol. 1 of *The Feynman Lectures on Physics.* Reading, Mass.: Addison-Wesley, 1963, pp.1–2.

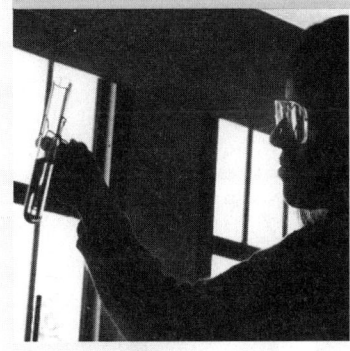

The Science Content Standards for Grades Nine Through Twelve

The Science Content Standards for Grades Nine Through Twelve

The science content standards for kindergarten through grade eight provide the background for students to succeed with the science content standards for grades nine through twelve. Aligning the high school curriculum to offer standards-based courses for every student will put new demands on schools and science departments. However, the reward for successfully meeting the challenge will be that high school graduates can attain the highest level of science literacy achieved by students in more than two decades.

Changing to a program based on the science content standards will require a restructuring of the high school curriculum, although the science that was generally taught in California before the *Science Content Standards for California Public Schools* was published is mostly included in the standards.[1] The successful implementation of standards-based kindergarten through grade eight programs aligned to this *Science Framework* should enable more students to take standards-based courses in high school. This chapter provides guidance for teaching students who have mastered the kindergarten through grade eight materials. To achieve this mastery will require many years of effort, and school districts should adjust their programs appropriately as their students have the opportunity to learn the prerequisite material in the earlier grades.

School districts are responsible for their curriculum and must decide how to structure their courses to teach the science standards. Traditionally, biology has been taught in the tenth grade, followed by chemistry and then possibly by physics. However, this sequence dates from a time when the content of the biology course was descriptive and that of the physics course was the most quantitative among the science disciplines. The high school science standards allow for other structures. Because districts need flexibility to design their own course structure, this chapter is presented in modular format—no sequence or emphasis is prescribed.

Appropriate to the rigor of the standards, each section covers a particular scientific discipline: physics, chemistry, biology/life sciences, and earth sciences. Along with meeting the subject-matter requirements for science, every student should learn the content in the full set of Investigation and Experimentation standards and have an opportunity to learn the slightly more advanced material in the standards that are marked with an asterisk.

In 1997 California established the Digital High School program, ensuring that all high schools throughout the state would have access to technology to improve student achievement in science and other academic subjects. Many schools purchased materials for scientific-based technology, and their use should be integrated into science programs. Technology can be used to teach some science standards and to assess students' understanding. Science education provides an opportunity to instruct students in gathering, graphing, tracking, and interpreting data through the use of technological tools, such as word processing, spreadsheets, and database development. Related concepts from science, mathematics, and language arts can be merged in the development of a science experiment and its subsequent analysis.

Safety is always the foremost consideration in the design of demonstrations, laboratories, and science experiments. The importance of safety is evident because scientists and engineers in universities and industries are required to follow strict health and safety regulations. Safety needs to be taught. Teachers should be familiar with the *Science Safety Handbook for California Public Schools*.[2] It contains specific, useful information relevant to classroom science teachers. School administrators, teachers, parents/guardians, and students have a legal and moral obligation to promote safety in science education. Knowing and following safe practices in science are a part of understanding the nature of science and scientific enterprise.

Physics

Many scientists and engineers consider physics the most basic of all sciences. It covers the study of motion, forces, energy, heat, waves, light, electricity, and magnetism. Physics focuses on the development of models deeply rooted in scientific inquiry, in which mathematics is used to describe and predict natural phenomena and to express principles and theories. Understanding physics requires the ability to use algebra, geometry, and trigonometry. This need for mathematics has kept all but a very few students in this country from studying physics. Other countries, however, have met this challenge by introducing the concepts of physics to students during a period of several years, starting in the earlier grade levels. Topics requiring little or no mathematics are introduced first, and students progress to more sophisticated and quantitative treatments as they learn more mathematics. The California standards emulate this successful approach.

All students can learn high school physics. Many will have enough foundational skills and knowledge of mathematics from their science curriculum in kindergarten through grade eight to study motion, forces, heat, and light. In high school, students should develop a working knowledge of algebra, geometry, and simple trigonometry to understand and gain access to the power of physics. Some will need to learn or relearn algebra, geometry, and trigonometry skills while studying physics. The need for such mathematics review should lessen over time as California's rigorous mathematics standards are implemented. Students who intend to pursue careers in science or engineering will need to master the physics content called for in the California standards, including the standards marked with an asterisk. (Note that equations appearing in this section are numbered consecutively.)

STANDARD SET 1. Motion and Forces

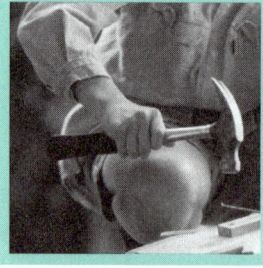

Motion deals with the changes of an object's position over time. Inherent in any useful study of motion is the concept of force, which represents the existence of physical interactions. Although Newton's laws provide a good platform from which to analyze forces, those laws do not address the origin of forces. Fundamental forces in nature govern the physical behavior of the universe. One of these fundamental forces, gravity, influences objects with mass but acts at a distance, or without any direct contact between the objects. The electromagnetic force is also a fundamental force that operates across a distance. These standards on motion and forces provide the foundation for understanding some key similarities—and differences—between these two forces. A working knowledge of basic algebra and geometry is an essential prerequisite for studying these concepts.

In standard sets presented earlier at lower grade levels, students were introduced to the idea that the motion of objects can be observed and measured, and they learned that a force can change the motion of an object by giving it a push or a pull. The topic of "Motion and Forces" at the high school level builds directly on the eighth grade Standard Set 1, "Motion," and Standard Set 2, "Forces," both of which introduce the notions of balanced forces and of net force (see Chapter 4). Students should know the difference between speed and velocity and should be able to interpret graphs for linear motion that plot relationships between two variables, such as speed versus time. Students should also understand the vector nature of forces. The concepts of gravity and of inertia as a resistance to a change in motion should have been introduced in the eighth grade.

> **1. Newton's laws predict the motion of most objects. As a basis for understanding this concept:**
>
> **a.** *Students know* how to solve problems that involve constant speed and average speed.

The rate at which an object moves is called its *speed*. Speed is measured in distance per unit time (e.g., meters/second). Velocity **v** is a vector quantity and therefore has both a magnitude—the speed—and a direction. If an object travels at a constant speed, a simple linear relationship exists between the speed, or rate of motion r; distance traveled d; and time t, as shown in

$$d = rt. \qquad (eq.\ 1)$$

If speed does not remain constant but varies with time, *average speed* can be defined as the total distance traveled divided by the total time required for the trip.

> **1. b.** *Students know* that when forces are balanced, no acceleration occurs; thus an object continues to move at a constant speed or stays at rest (Newton's first law).

If an object's velocity v changes with time t, then the object is said to accelerate. For motion in one dimension, the definition of acceleration a is

$$a = \Delta v/\Delta t, \qquad (eq.\ 2)$$

where the Greek capital letter delta (Δ) stands for "a change of." *Acceleration* is defined as change in velocity per unit time. (Another way to state this definition is that *acceleration* is a change in distance per unit time per unit time, producing acceleration units of, for example, m/s² [meters per second squared or meters per second per second].) Acceleration is a vector quantity and therefore has both magnitude and direction. A push or a pull (force) needs to be applied to make an object accelerate. Force is another vector quantity.

A vector quantity, such as force, can be resolved into its x, y, and z components, F_x, F_y, and F_z. More than one force can be applied to an object simultaneously. If the forces point in the same direction, their magnitudes add; if the forces point in opposite directions, their magnitudes subtract. The net (overall) force can be calculated by adding forces along a line algebraically and keeping track of the direction and signs. If an object is subject to only one force, or to multiple forces whose vector sum is not zero, there must be a net force on the object. However, if there is no net force on an object already in motion, that object continues to move at a constant velocity. An object at rest remains at rest if no net force is applied to it. This principle is Newton's first law of motion.

> **1. c.** *Students know* how to apply the law *F = ma* to solve one-dimensional motion problems that involve constant forces (Newton's second law).

If a net force is applied to an object, the object will accelerate. The relationship between the net force *F* applied to an object, the object's mass *m,* and the resulting acceleration *a* is given by Newton's second law of motion

$$F = ma \,. \tag{eq. 3}$$

If mass is in kilograms (kg) and acceleration is in meters per second squared (m/s^2), then force is measured in Newtons, with 1 Newton = 1 kilogram-meter per second squared (1 kg-m/s^2).

If the net force on an object is constant, then the object will undergo constant acceleration. When studying constant force, students should be able to make use of the following equations to describe the motion of an object in one dimension at any elapsed time *t* by calculating its velocity *v* and distance from the origin *d:*

$$v = v_0 + at \,, \tag{eq. 4}$$

$$d = d_0 + v_0 t + \tfrac{1}{2} a t^2 \,. \tag{eq. 5}$$

In these equations *m* is the mass, v_0 is the initial velocity, d_0 is the initial position (distance from origin) of the object, and *t* is the time during which the force *F* is applied.

> **1. d.** *Students know* that when one object exerts a force on a second object, the second object always exerts a force of equal magnitude and in the opposite direction (Newton's third law).

Newton's third law of motion is more commonly stated as, "To every action there is always an equal and opposite reaction." The mutual reactions of two bodies are always equal and point in opposite directions. Mathematically stated, if object 1 pushes on object 2 with a force $\mathbf{F}_{12}$, then object 2 pushes on object 1 with a force $\mathbf{F}_{21}$ such that

$$\mathbf{F}_{21} = -\mathbf{F}_{12} \,. \tag{eq. 6}$$

This universal law applies, for example, to every object on the surface of Earth. Trees, rocks, buildings, and cars, even the atmosphere, are all subject to the downward force of gravity. In all cases Earth exerts an equal and opposite upward push on the objects. Stars exist because of the balance between the inward force of gravity and the outward pressure of their hot interior gases.

1. e. *Students know* the relationship between the universal law of gravitation and the effect of gravity on an object at the surface of Earth. (See Standard 1.m.*)

Since the time of Galileo's reputed experiment of dropping objects from the tower of Pisa, it has been understood that in the absence of air resistance, all objects near Earth's surface, regardless of their mass or composition, accelerate downward toward Earth's center at 9.8 m/s². Through Newton's second law, this principle can be expressed as

$$F = w = mg \text{ (where g} \approx 9.8 \text{m/s}^2 \text{ is the acceleration due to gravity).} \quad \text{(eq. 7)}$$

The gravitational force pulling on an object is called the object's weight w and is measured in Newtons.

1. f. *Students know* applying a force to an object perpendicular to the direction of its motion causes the object to change direction but not speed (e.g., Earth's gravitational force causes a satellite in a circular orbit to change direction but not speed).

A force that acts on an object may act in any direction. The component of the force parallel to the direction of motion changes the speed of the object, and the components perpendicular to the motion change the direction in which the object travels.

1. g. *Students know* circular motion requires the application of a constant force directed toward the center of the circle.

An object moving with constant speed in a circle is in uniform circular motion. The direction of motion continuously changes because of a force that always points inward toward the center of the circle. Such a centrally directed force is called a *centripetal force*. If the mass of the object is m, its speed is v, and the radius of the circle in which the object travels is r, then the magnitude of the force causing the circular motion is

$$F_c = mv^2/r. \quad \text{(eq. 8)}$$

Examples of centripetal forces are the tension in a string attached to a ball that is swung in a circle, the pull of gravity on a satellite in orbit around Earth, the electrical forces that deflect electrons in a television tube, and the magnetic forces that turn a charged particle.

1. h.* *Students know* Newton's laws are not exact but provide very good approximations unless an object is moving close to the speed of light or is small enough that quantum effects are important.

Newton's laws are not exact but are excellent approximations valid in domains involving low speeds and macroscopic objects. However, when the speed of an object approaches the speed of light (3×10^8 m/s), Einstein's theory of special relativity is required to describe the motion of the object accurately. Among the major differences between Einstein's and Newton's theories of mechanics are that (1) the maximum attainable speed of an object is the speed of light; (2) a moving clock runs more slowly than does a stationary one; (3) the length of an object depends on its velocity with respect to the observer; and (4) the apparent mass of an object increases as its speed increases.

The other domain in which Newtonian mechanics breaks down is that of very small objects, such as atoms or atomic nuclei. Here the wavelike nature of matter becomes important, and quantum mechanics better describes the submicroscopic world. Newtonian mechanics assumes that if the motion of a particle is measured with great accuracy and all the masses and forces that are involved are also known, it is always possible to predict with equally great accuracy the future state of motion of the particle. Quantum mechanics shows that such certainty is not always possible. Sometimes only the probability of an outcome can be predicted.

> **1. i.*** *Students know* how to solve two-dimensional trajectory problems.

Students can consider the problem of a ball of mass m thrown upward into the air at some angle. The motion of the ball will have horizontal and vertical components that are independent of one another. If air resistance is ignored, there will be no horizontal force acting against the ball to slow it down. While the ball is in flight then, only a single vertical force, gravity, is acting on the ball (e.g., $F = w = mg$ downward). If students know the angle and the height from which the ball is thrown and the ball's initial velocity, they will be able to predict the path of the ball and to calculate how high the ball will go, how far it will travel before it strikes the ground, and how long it will be in the air.

> **1. j.*** *Students know* how to resolve two-dimensional vectors into their components and calculate the magnitude and direction of a vector from its components.

In a two-dimensional system, two quantities are needed to describe a vector. A vector **r** can be completely specified by a magnitude r and an angle Φ or by its x and y components (i.e., r_x and r_y). Simple trigonometry can be applied to resolve a vector into its components (e.g., $r_x = r \cos \Phi$ and $r_y = r \sin \Phi$) and to calculate the magnitude and direction of a vector from its components ($r^2 = r_x^2 + r_y^2$ and $\tan \Phi = r_y/r_x$).

> **1. k.*** *Students know* how to solve two-dimensional problems involving balanced forces (statics).

A body at rest that is subject to no net force is in static equilibrium. Examples of static equilibrium are a book resting on the surface of a table and a ladder leaning at rest against a wall. Because the book and table remain at rest does not imply that no forces act on these objects but does imply that the vector sum of all these forces is zero. In particular, the components of the forces in any particular direction sum to zero. Thus for an object that remains at rest,

$$\sum F_y = 0, \qquad \text{(eq. 9)}$$

where the Greek capital letter sigma (Σ) means to "sum over or add" and F_y represents the components in any chosen direction y of the forces acting on the object. One sample problem appears in Figure 2, "Calculation of Force." Students are given the weight of a hanging object, the lengths of the ropes holding it in place, and the distance between the anchors. The students are asked to calculate the forces, called *tension,* along ropes of equal length. Students find this problem difficult because the vector force diagram they should use to solve the problem is often confused with the physical lengths of the ropes.

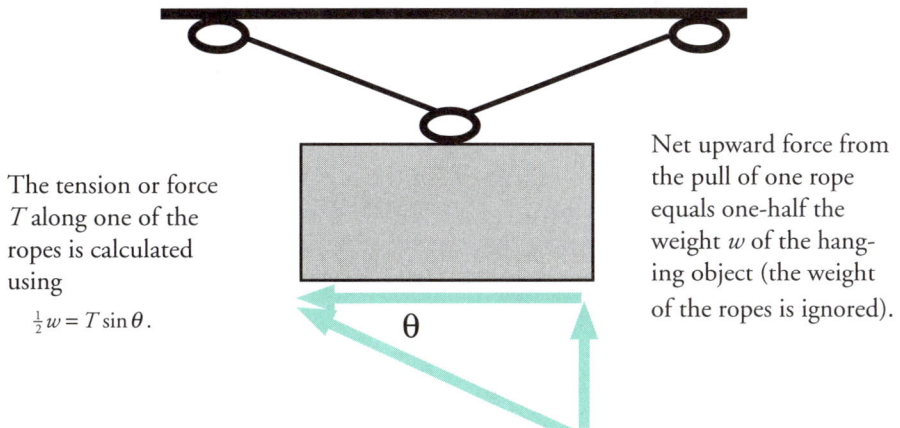

Fig. 2. Calculation of Force

> **1. l.*** *Students know* how to solve problems in circular motion by using the formula for centripetal acceleration in the following form: $a = v^2/r$.

The speed of an object undergoing uniform circular motion does not vary, but the object's direction does and hence the object's velocity. Thus the object is constantly accelerating. The magnitude of this centripetal acceleration is

$$a_c = F_c/m = v^2/r, \qquad \text{(eq. 10)}$$

and the direction of the centripetal acceleration vector rotates so that it always points inward toward the center of the circle.

> **1. m.*** *Students know* how to solve problems involving the forces between two electric charges at a distance (Coulomb's law) or the forces between two masses at a distance (universal gravitation).

Standard Set 5 for physics, "Electric and Magnetic Phenomena," which appears later in this section, shows that the origin of the force between two masses and between two electric charges is entirely different. However, the forces involved, the gravitational and the electromagnetic forces, are both inverse square relationships. Coulomb's law (in a vacuum) is written

$$F_q = kq_1q_2/r^2 , \qquad \text{(eq. 11)}$$

where $k = 9 \times 10^9$ Nm²/coul², q_1 and q_2 are charges (positive [+] or negative [−]), r is the distance separating the charges, and F_q is the force resulting from the two charges. The force is repulsive if the charges are the same sign and attractive if they are different.

Newton's law of universal gravitation states that if two objects have masses m_1 and m_2, with centers of mass separated from each other by a distance r, then each object exerts an attractive force on the other; the magnitude of this force is

$$F_g = Gm_1m_2/r^2 , \qquad \text{(eq. 12)}$$

where G is the universal gravitational constant, equal to 6.67×10^{-11} newton-m²/kg². For the case of a small object falling freely near the surface of Earth, students should understand that

$$g = Gm_e/r_e^2 = 9.8 \text{ m/s}^2 , \qquad \text{(eq. 13)}$$

where m_e and r_e are the mass and radius of Earth. Students might be interested to know that Henry Cavendish's measurement of G, completed around the year 1800, was the last piece of information needed to calculate the mass of Earth.

STANDARD SET 2. Conservation of Energy and Momentum

The concept of energy was introduced and discussed several times in the lower grades, from the physical sciences through the life sciences. In fact, every process involves some transfer of energy. In Standard Set 2 *energy* is classified as *kinetic,* meaning related to an object's motion, or as *potential,* meaning related to an object's stored energy. The energy of a closed system is conserved. Another useful conservation law, conservation of momentum, is introduced and is shown to be a direct consequence of Newton's laws. The power and importance of these conservation laws are that they allow physicists to predict the motion of objects without having to know the details of the dynamics and interactions in a given system.

Through the standard sets introduced in the lower grade levels, students should have learned about forces and motion and the idea of energy. They should have been taught the role of energy in living organisms and the effects of energy on Earth's weather. The standards presented earlier also call for student exposure to energy conservation, a concept that is essential to the topics contained in the high school physics standard sets 3, 4, and 5 and in several standard sets in chemistry and earth sciences.

> **2. The laws of conservation of energy and momentum provide a way to predict and describe the movement of objects. As a basis for understanding this concept:**
>
> **a.** *Students know* how to calculate kinetic energy by using the formula $E = \frac{1}{2}mv^2$.

Kinetic energy is energy of motion. The kinetic energy of an object equals the work that was needed to create the observed motion. This work can be related to the net force applied to the object along the line of the motion. The work done on an object by a force is equal to the component of the force along the direction of motion multiplied by the distance the object moved:

$$W = Fd. \qquad \text{(eq. 14)}$$

The work needed to accelerate an object of mass m from rest to a speed v is $\frac{1}{2}mv^2$. This quantity is defined as the kinetic energy E. The units of energy are joules, in which 1 joule = 1 kilogram-meter squared per second squared (1 kg-m²/s²) = 1 newton-meter. Energy is a *scalar* quantity, meaning that energy has a magnitude but no direction.

> **2. b.** *Students know* how to calculate changes in gravitational potential energy near Earth by using the formula (change in potential energy) = mgh (h is the change in the elevation).

Students can combine equations (3) and (14) to find the work done in lifting an object of weight mg through a vertical distance h, as shown in

$$W = mgh. \qquad \text{(eq. 15)}$$

Work and energy have the same units. Therefore, one can define mgh as the change in gravitational potential energy associated with the change in elevation h of the mass m.

> **2. c.** *Students know* how to solve problems involving conservation of energy in simple systems, such as falling objects.

Equations (4) and (5) can be used to show that if the object dealt with in Standard 2.b is released from rest and allowed to fall freely, it will strike the ground with a speed

$$v = \sqrt{2gh}, \qquad \text{(eq. 16)}$$

and its kinetic energy at the instant of impact will be

$$E = \tfrac{1}{2}mv^2 = \tfrac{1}{2}m(2gh) = mgh. \qquad \text{(eq. 17)}$$

The total energy T of the object is then defined as the sum of kinetic plus potential energy

$$T = E + PE. \qquad \text{(eq. 18)}$$

This sum is conserved in a closed system for such forces as gravity and electromagnetic interactions and those produced by ideal springs. Thus,

$$\Delta E + \Delta PE = 0. \qquad \text{(eq. 19)}$$

Therefore, the change in kinetic energy equals the negative of the change in potential energy. This principle is a consequence of the law of the conservation of energy. Energy can be converted from one form to another, but in a closed system the total energy remains the same.

> **2. d.** *Students know* how to calculate momentum as the product *mv*.

The momentum **p** of an object is defined as the product of its mass *m* and its velocity **v**. Momentum is thus a vector quantity, having both a magnitude and a direction. The units of momentum are kg-m/s. The magnitude of the momentum is *mv,* the product of the object's mass and its speed.

> **2. e.** *Students know* momentum is a separately conserved quantity different from energy.

If no net force is acting on an object or on a system of objects, the momentum remains constant. That is, neither its magnitude nor its direction changes with time. Conservation of momentum is another fundamental law of physics.

> **2. f.** *Students know* an unbalanced force on an object produces a change in its momentum.

As discussed in the section for Standard 1.c, if the net force on an object is not zero, then its velocity and hence its momentum will change. Motion resulting from a constant force **F** acting on an object for a time Δt causes a change in momentum of $\mathbf{F}\Delta t$. This change in momentum is called an *impulse*. (Note that the units of impulse are the same as those of momentum [i.e., newton-second = kg-m/s].) Depending on the direction of the force, the impulse can increase, decrease, or change the direction of the momentum of an object.

> **2. g.** *Students know* how to solve problems involving elastic and inelastic collisions in one dimension by using the principles of conservation of momentum and energy.

Momentum is always conserved in collisions. Collisions that also conserve kinetic energy are called *elastic collisions;* that is, the kinetic energy before and after the collision is the same. Billiard balls colliding on smooth pool tables and gliders colliding on frictionless air tracks are approximate examples. Collisions in which kinetic energy is not conserved are called *inelastic collisions.* An example is a golf ball colliding with a ball of putty and the two balls sticking together. Some of the kinetic energy in inelastic collisions is transformed into other types of energy, such as thermal or potential energy. In all cases the total energy of the system is conserved.

> **2. h.*** *Students know* how to solve problems involving conservation of energy in simple systems with various sources of potential energy, such as capacitors and springs.

An ideal spring is an example of a conservative system. The force required either to stretch or to compress a spring by a displacement x from its equilibrium (unstretched) length is

$$F = kx, \qquad \text{(eq. 20)}$$

where k is the spring constant that measures a spring's stiffness. A graph of the magnitude of this force as a function of the compression shows that the force varies linearly from zero to kx as the spring is compressed. The area under this graph is the work done in compressing the spring and is equal to

$$\tfrac{1}{2}(\text{base})(\text{height}) = \tfrac{1}{2}kx^2. \qquad \text{(eq. 21)}$$

This is also the potential energy stored in the spring.

A capacitor stores charge. The charge Q that is stored depends on the voltage V according to

$$Q = CV, \qquad \text{(eq. 22)}$$

where the constant C is called the *capacitance*. (Notice that this equation and the equation for a spring [eq. 20] have the same form.) The energy stored in a capacitor is given by the equation

$$E = \tfrac{1}{2}CV^2, \qquad \text{(eq. 23)}$$

which also has the same form as the equation that gives the energy stored by a spring.

STANDARD SET 3. Heat and Thermodynamics

The concept of heat (thermal energy) is related to all scientific disciplines. Energy transfer, molecular motion, temperature, pressure, and thermal conductivity are integral parts of physics, chemistry, biology, and earth science. Thermodynamics deals with exchanges of energy between systems.

If students in high school have not yet covered the chemistry standards, the related topics from those standards should be introduced. (See the following standards for chemistry in this chapter: 4.a through 4.h, "Gases and Their Properties," and 7.a through 7.d, "Chemical Thermodynamics." Specific chemistry topics that are useful or necessary for promoting a more complete understanding of Standard

Set 3 are specifically mentioned, when relevant, under the sections with detailed descriptions.

At the atomic and molecular levels, all matter is continuously in motion. For example, individual molecules of nitrogen, oxygen, and other gases that make up the air inside a balloon move at varying speeds in random directions, vibrating and rotating. The collisions of these molecules with the inner surface of the balloon create the pressure that supports the balloon against atmospheric pressure.

Considerable confusion exists in scientific literature about the definitions of the terms *heat* and *thermal energy*. Some texts define *heat* strictly as "transfer of energy." These science content standards use the term *heat* interchangeably with *thermal energy*. However, it is less confusing to reserve the term *heat* for thermodynamic situations in which energy is transferred either because of differences in temperature or through work done by or on a system. In this sense both *heat* and *work* have meaning only as they describe energy exchanges into and out of the system, adding or subtracting from a system's store of internal energy.

Students, just like scientists of the eighteenth century, might easily fall prey to the misconception that heat is a substance. Students should be cautioned that heat is energy, not a material substance, and that *heat flow* refers not to material flow but to the transfer of energy from one place to another. Confusion is most apt to arise when dealing with heat transfer by convection; that is, when heat is transferred through actual motion of hot and cold material along a thermal gradient. Heating a material such as air causes it to expand and leads to differences in density that drive the movement of heated material.

Students also often confuse temperature and heat. From a molecular viewpoint, *temperature* is a measure of the average translational kinetic energy of a molecule, as shown in equation (27). (See also Standard Set 7 in the chemistry section in this chapter.) Studies of the temperature of materials as they pass through phase transitions may also help students understand the differences and relationships between heat and temperature.

A way to avoid confusion is to reserve the use of the word *heat* for situations in which heat transfer is involved, as described in the next section.

> **3. Energy cannot be created or destroyed, although in many processes energy is transferred to the environment as heat. As a basis for understanding this concept:**
>
>   **a.** *Students know* heat flow and work are two forms of energy transfer between systems.

Heat transfer is energy flow from one system to another because of differences in temperature or because of mechanical work. The energy that flows into a pot of cold water put on a hot stove is an example of heat transfer. This energy increases the kinetic energy of the random motion of the molecules of water and therefore the temperature of the water rises. When the water reaches 100°C, a new phenomenon, a phase transition, occurs: the water vaporizes, or boils. Although energy continues to flow into the water, the kinetic energy of the water molecules does not increase; therefore the temperature of the water remains constant. As the water changes from a liquid to a gas, the energy goes instead into breaking the bonds that hold one molecule of liquid water to another. The energy required (per unit mass or mole of liquid) to change a particular liquid at its boiling temperature into a gas is called the liquid's *latent heat of vaporization.*

Mechanical work can change temperature too (e.g., when the forces of friction heat objects or when a gas is compressed and so warms). Conversely, changes in temperature can do mechanical work (e.g., warming a container of gas that is sealed by a piston will cause the gas to expand and the piston to move).

Heat is energy that moves between a system and its environment because of a temperature difference between them. Every system has its internal energy, that is, the energy required to assemble the system; and this energy is independent of any particular path or means by which the system is assembled. The transfer of internal energy from one system to another, because of a temperature difference, is known as *heat flow.* There are three basic kinds of heat flow: conduction, convection, and radiation. Students should have first learned about these processes in the sixth grade.

As heat is transferred to a system (object), the temperature of the system (object) may increase. Substances vary in the amount of heat necessary to raise their temperatures by a given amount. More mass in the system clearly requires more heat for a given temperature change. An expression that illustrates the relationship between the amount of heat transferred and the corresponding temperature change is shown in equation (24). The change in temperature ΔT is proportional to the amount of heat added. This relationship is specified by

$$Q = mC\Delta T, \qquad \text{(eq. 24)}$$

where Q is the internal energy added by heat transfer to the system from the surroundings, ΔT is the difference in temperature between the final and initial states of the system, m is the system's mass, and C is the specific heat of the substance (in joules/gram-°C or calories/gram-°C). *Specific heat* is a characteristic property of a material. The unit of specific heat is energy divided by mass and temperature change (e.g., calories/gram-degree).

Water, which serves as a standard against which all other materials may be compared, has a specific heat of one calorie/gram-degree. In other words, one calorie of heat is required to raise one gram of water one degree Celsius. When a gram of water cools one degree, one calorie is liberated. This value is large compared with those of other substances. Therefore, it takes much more heat to warm water than it does to raise the temperature of the same amount of most other substances. This fact has important implications for weather and climate and is one reason the weather is "tempered" in coastal areas (e.g., summers are cooler and winters are warmer than they are in inland areas at a similar latitude).

Equation (24) makes the distinction between heat and temperature quite clear. It specifies that heat can flow in or out of a system because of temperature difference alone. There are, however, other situations in which the addition or removal of heat is not accompanied by changes in temperature. These situations occur when a substance undergoes a change of phase, or state, such as when water evaporates or freezes. During phase changes, the absorption or release of heat takes place while the system remains at a constant temperature. For example, when ice melts in a glass of water that is sufficiently well mixed, the temperature of the water remains at the freezing point of water. Additional heating of the water raises its temperature only after the ice has melted.

> **3. b.** *Students know* that the work done by a heat engine that is working in a cycle is the difference between the heat flow into the engine at high temperature and the heat flow out at a lower temperature (first law of thermodynamics) and that this is an example of the law of conservation of energy.

The total energy of an isolated system is the sum of the kinetic, potential, and thermal energies. A system is isolated when the boundary between the system and the surroundings is clearly defined. Total energy is conserved in all classical processes. Thus, the law of conservation of energy can be restated as the first law of thermodynamics; that is, for a closed system the change in the internal energy ΔU is given by the expression

$$\Delta U = Q - W, \qquad \text{(eq. 25)}$$

where Q is the internal energy added by heat transfer to the system from the surroundings and W is the work done by the system. The quantities ΔU, Q, and W in equation (25) can be negative or positive, depending on whether energy is converted from mechanical form into heat, as when work is done on the system, or on whether heat is transformed into mechanical energy, as when the system is doing work. By convention, Q is positive for heat added to the system and negative for heat transferred to the surroundings, and W is positive for work done by the system and negative for work done on the system. As a practical matter, energy that cannot be obtained as work is considered a loss to the system. Thus, the first law of thermodynamics indicates how much energy is available to do work.

A heat engine is a device for getting useful mechanical work from thermal energy. While part of the input heat energy Q_H, sometimes known as *heat of combustion,* is converted into useful work W, the remaining heat is lost to the environment as exhaust heat Q_L. That is, the work done by a heat engine is the difference between thermal energy flowing in at higher temperature and heat flowing out at lower temperature, as shown in the following equation:

$$W = Q_H - Q_L. \qquad \text{(eq. 26)}$$

This simple relationship is valid for an idealized engine, also called a *Carnot engine.*

> **3. c.** *Students know* the internal energy of an object includes the energy of random motion of the object's atoms and molecules, often referred to as *thermal energy*. The greater the temperature of the object, the greater the energy of motion of the atoms and molecules that make up the object.

The internal energy of objects is in the motion of their atoms and molecules and in the energy of the electrons in the atoms. For ideal gases, nearly realized by air molecules, heat transferred to the gas increases the average speed of the gas molecules. The higher the temperature, the greater the average speed. If it were possible to observe the motion of molecules in a gas at a fixed temperature, one would see molecules with different masses moving on average at different speeds. More massive molecules, for example, move more slowly because the average kinetic energy of each type of molecule is the same in the gas, and the kinetic energy is proportional to the product of the mass and the square of the velocity of the gas molecules. The pressure of a gas results from individual molecules bumping against containing walls and other objects. Each hit and change of direction causes a change in momentum and therefore a net force or push on the object hit. One molecule's contribution to total pressure is very small, but measurable pressures result when large numbers of fast-moving atoms or molecules participate in these collisions.

For an ideal gas system at thermal equilibrium, the kinetic energy of an individual gas molecule averaged over time is

$$E = \tfrac{3}{2}kT, \qquad (eq.\ 27)$$

where $k = 1.38 \times 10^{-23}$ joule/K, and T is the absolute temperature in Kelvin (K). The Kelvin temperature scale and its conversion to the customary Fahrenheit and Celsius scales are discussed in standards 4.d and 4.e in the chemistry section in this chapter.

> **3. d.** *Students know* that most processes tend to decrease the order of a system over time and that energy levels are eventually distributed uniformly.

Energy in the form of heat transfers from hot to cold, but not from cold to hot, regardless of whether that energy transfers by radiation, conduction, or convection. Why? Matter exists in discrete energy states (or levels). For tiny objects, such as a single electron, the difference in energy between one state and the next is big enough to be detected and measured. For the larger objects of everyday experience, such as a pebble, the difference is too small to detect; still, the discrete states exist, and it makes sense to speak of the probability with which any given system is to be found in any one of its possible states.

A system of many components has many states of given total energy because some components can have a larger fraction of that energy if others have less. Such a system evolves so that all states with the same total energy become equally probable. Heat flows from hot to cold because states in which components share energy equally vastly outnumber states in which they do not. A copper bar with one end hot and the other cold has many atoms with more kinetic energy on the hot end and many atoms with less kinetic energy on the cold end. Later, however, because the kinetic energy has been transferred from the hot end of the bar to the cold end, all the atoms will have nearly the same kinetic energy. The change can be interpreted as heat flowing from hot to cold until the temperature of the bar is uniform. Similarly, most physical processes disorder a system because disordered states vastly outnumber ordered ones. A drop of perfume evaporates because states in which molecules of perfume are scattered throughout a large volume of air vastly outnumber states in which the molecules are confined in the tiny volume of a drop.

> **3. e.** *Students know* that entropy is a quantity that measures the order or disorder of a system and that this quantity is larger for a more disordered system.

Students know from Standard 3.d that energy transferred as heat leads to the redistribution of energy among energy levels in the substances that compose the system. This redistribution increases the disorder of material substances. A quantity called *entropy* has been defined to track this process and to measure the randomness, or disorder, of a system. Entropy is larger for a disordered system than for an ordered one. Thus, a positive change in entropy, in which the final entropy is larger than the initial entropy, indicates decreasing order, also considered as increasing disorder. The properties of entropy fix the maximum efficiency with which energy stored as a temperature difference can be converted into work.

For a system at constant temperature, such as during melting or boiling, the change in entropy ΔS is given by

$$\Delta S = Q/T, \qquad \text{(eq. 28)}$$

where Q is the heat (thermal energy) that flows into or out of the system and T is the absolute temperature. The units of entropy are joules/K. All processes that require energy, for example, biochemical reactions that support life, occur only because the entropy increases as a result of the process.

> **3. f.*** *Students know* the statement "Entropy tends to increase" is a law of statistical probability that governs all closed systems (second law of thermodynamics).

The second law of thermodynamics states that all spontaneous processes lead to a state of greater disorder. When an ice cube melts and the water around it becomes cooler, for example, the internal energy of the ice-water system becomes more uniformly spread, or more disordered. Most processes in nature are irreversible because they move toward a state of greater disorder. A broken egg, for instance, is almost impossible to restore to its original ordered state.

Another statement of the second law of thermodynamics is that in a closed system all states tend to become equally probable. Calculating the statistical probability of a condition involves counting all the ways to distribute energy in a system, and that procedure involves mathematics that is more complex than most students will have mastered. However, most students can recognize that there are many more ways to distribute energy approximately evenly within a system than there are ways to have energy concentrated. As spontaneous processes make all ways equally probable, a system thus becomes more likely to be found with its energy distributed than concentrated, and so the system becomes disordered.

Students who complete these standards will have learned the first and second laws of thermodynamics. They should understand that when physical change occurs, energy must be conserved, and some of this energy cannot be recovered for useful work because it has added to the disorder of the universe.

> **3. g.*** *Students know* how to solve problems involving heat flow, work, and efficiency in a heat engine and know that all real engines lose some heat to their surroundings.

As implied in Standard 3.b, when heat flows from a body at high temperature to one at low temperature, some of the heat can be transformed into mechanical work. This principle is the basic concept of the heat engine. The remainder of the heat is transferred to the surroundings and therefore is no longer available to the system to do work. This transferred heat is never zero; therefore, some heat must always be transferred to the surroundings. Examples of practical heat engines are steam engines and internal combustion engines. Steam at a high temperature T_H pushes on a piston or on a turbine and does work. Steam at a lower temperature T_L is then drawn off from the engine into the air. When an idealized (i.e., reversible) engine completes a cycle, the change in entropy is zero. Equation (28) shows that

$$Q_H/T_H = Q_L/T_L. \qquad \text{(eq. 29)}$$

When this relation is combined with the conservation law of equation (26), the maximum possible efficiency, denoted as "eff," can be calculated as

$$\text{eff}(\%) = 100 \times W/Q_H = 100 \times (T_H - T_L)/T_H, \qquad \text{(eq. 30)}$$

where efficiency is the ratio of work done by the engine to the heat supplied to the engine. The efficiency of converting heat to work is proportional to the difference between the high and low temperatures of the engine's working fluids, usually gases. For a Carnot engine to be 100 percent efficient, the temperature of the exhaust heat needs to be absolute zero, an impossible occurrence.

STANDARD SET 4. Waves

Students can be introduced to this standard set by learning to distinguish between mechanical and electromagnetic waves. In general, a *wave* is defined as the propagation of a disturbance. The nature of the disturbance may be mechanical or electromagnetic. Mechanical waves, such as ocean waves, acoustic waves, seismic waves, and the waves that ripple down a flag stretched taut by a wind, require a medium for their propagation and gradually lose energy to that medium as they travel. Electromagnetic waves can travel in a vacuum and lose no energy even over great distances. When electromagnetic waves travel through a medium, they lose energy by absorption, a phenomenon that explains why light signals sent through the most transparent of optical fibers still need to be amplified and repeated. In contrast, light emitted from distant galaxies has traveled great distances without the aid of amplification, an indication that a relatively small amount of material is in the light's path.

Waves transfer energy from one place to another without net circulation or displacement of matter. Light, sound, and heat energy can be transmitted by waves across distances measured from fractions of a centimeter to many millions of kilometers. Exertion of a direct mechanical force, such as a push or a pull, on a physical body is an example of energy transfer by direct contact. However, for transfer to occur, objects do not need to be in direct physical contact with a source of energy. For instance, light transmits from a distant star, heat radiates from a fire, and sound propagates from distant thunder. Energy may be transferred by radiation, for example, from the Sun to Earth; therefore, radiation is also an example of a noncontact energy transfer. Both sight and hearing are senses that can perceive energy patterned to convey information without direct contact between the source and the sensing organ.

If students take physics before they have studied other high school science courses, the teacher may find it useful to cross-reference materials on pressure, heat, and solar radiation from the following standards in this chapter: 4.a and 7.a in the chemistry section and 4.a through 4.c in the earth sciences section. Algebra, geometry, and simple trigonometric skills are required for some of the advanced topics in this standard set. Students with a good foundation in algebra and geometry can be taught the trigonometry necessary to solve problems in this standard set.

> **4. Waves have characteristic properties that do not depend on the type of wave. As a basis for understanding this concept:**
>
> **a.** *Students know* waves carry energy from one place to another.

Waves may transport energy through a vacuum or through matter. Light waves, for example, transport energy in both fashions, but sound waves and most other waves occur only in matter. However, even waves propagating through matter transport energy without any net movement of the matter, thus differing from other means of energy transport, such as convection, a waterfall, or even a thrown object.

> **4. b.** *Students know* how to identify transverse and longitudinal waves in mechanical media, such as springs and ropes, and on the earth (seismic waves).

Waves that propagate in mechanical media are either longitudinal or transverse waves. The disturbance in longitudinal waves is parallel to the direction of propagation and causes compression and expansion (rarefaction) in the medium carrying the wave. The disturbance in transverse waves is perpendicular to the direction of propagation of the wave. Examples of longitudinal waves are sound waves and *P*-type earthquake waves. In transverse waves a conducting medium, or a test particle inserted in the wave, moves perpendicular to the direction in which the wave propagates. Examples of transverse waves are *S*-type earthquake waves and electromagnetic (or light) waves.

> **4. c.** *Students know* how to solve problems involving wavelength, frequency, and wave speed.

All waves have a velocity **v** (propagation speed and direction), a property that represents the rate at which the wave travels. Only periodic, sustained waves can be easily characterized through the properties of wavelength and frequency. However, most real waves are *composite,* meaning they can be understood as the sum of a few or of many waveforms, each with an amplitude, a wavelength, and a frequency.

Wavelength λ is the distance between any two repeating points on a periodic wave (e.g., between two successive crests or troughs in a transverse wave or between adjacent compressions or expansions [rarefactions] in a longitudinal wave). Wavelength is measured in units of length.

Frequency *f* is the number of wavelengths that pass any point in space per second. A wave will make any particle it encounters move in regular cycles, and frequency is also the number of such cycles made per second and is often abbreviated as cycles per second. The unit of frequency is the inverse second (s^{-1}), a unit also called the hertz (Hz).

Periodic wave characteristics are related to each other. For example,

$$v = f\lambda .$$ (eq. 31)

> **4. d.** *Students know* sound is a longitudinal wave whose speed depends on the properties of the medium in which it propagates.

Sound waves, sometimes called *acoustic waves,* are typically produced when a vibrating object is in contact with an elastic medium, which may be a solid, a liquid, or a gas. A sound wave is longitudinal, consisting of regions of high and low pressure (and therefore of compression and rarefaction) that propagate away from the source. (Note that sound cannot travel through a vacuum.) In perceiving sound, the human eardrum vibrates in response to the pattern of high and low pressure. This vibration is translated into a signal transmitted by the nervous system to the brain and interpreted by the brain as the familiar sensation of sound. Microphones similarly translate vibrations into electrical current. Sound speakers reverse the process and change electrical signals into vibrational motion, recreating sound waves.

An acoustic wave attenuates, or reduces in amplitude, with distance because the energy in the wave is typically spread over a spherical shell of ever-increasing area and because interparticle friction in the medium gradually transforms the wave's energy into heat. The speed of sound varies from one medium to another, depending primarily on the density and elastic properties of the medium. The speed of sound is typically greater in solid and liquid media than it is in gases.

> **4. e.** *Students know* radio waves, light, and X-rays are different wavelength bands in the spectrum of electromagnetic waves whose speed in a vacuum is approximately 3×10^8 m/s (186,000 miles/second).

Electromagnetic waves consist of changing electric and magnetic fields. Because these fields are always perpendicular to the direction in which a wave moves, an electromagnetic wave is a transverse wave. The electric and magnetic fields are also always perpendicular to each other. Concepts of electric and magnetic fields are introduced in Standard Set 5, "Electric and Magnetic Phenomena," in this section. The range of wavelengths for electromagnetic waves is very large, from less than nanometers (nm) for X-rays to more than kilometers for radio waves. The human eye senses only the narrow range of the electromagnetic spectrum from 400 nm to 700 nm. This range generates the sensation of the rainbow of colors from violet through the respective colors to red. In a vacuum all electromagnetic waves travel at the same speed of 3×10^8 m/s (or 186,000 miles per second). In a medium the speed of an electromagnetic wave depends on the medium's properties and on the frequency of the wave. The ratio of the speed of a wave of a given frequency in a vacuum to its speed in a medium is called that medium's *index of refraction.* For visible light in water, this number is approximately 1.33.

> **4. f.** *Students know* how to identify the characteristic properties of waves: interference (beats), diffraction, refraction, Doppler effect, and polarization.

A characteristic and unique property of waves is that two or more can occupy the same region of space at the same time. At a particular instant, the crest of one wave can overlap the crest of another, giving a larger displacement of the medium from its condition of equilibrium *(constructive interference);* or the crest of one wave can overlap the trough of another, giving a smaller displacement *(destructive interference).* The effect of two or more waves on a test particle is that the net force on the particle is the algebraic sum of the forces exerted by the various waves acting at that point.

If two overlapping waves traveling in opposite directions have the same frequency, the result is a standing wave. There is a persistent pattern of having no displacement in some places, called *nulls* or *nodes,* and large, oscillating displacements in others, called *maxima* or *antinodes.* If two overlapping waves have nearly the same frequency, a node will slowly change to a maximum and back to a node, and a maximum will slowly change to a node and back to a maximum. For sound waves this periodic change leads to audible, periodic changes from loud to soft, known as *beats.*

Diffraction describes the constructive and destructive patterns of waves created at the edges of objects. Diffraction can cause waves to bend around an obstacle or to spread as they pass through an aperture. The nature of the diffraction patterns of a wave interacting with an object depends on the ratio of the size of the obstacle to the wavelength. If this ratio is large, the shadows are nearly sharp; if it is small, the shadows may be fuzzy or not appear at all. Therefore, a hand can block a ray of light, whose average wavelength is about 500 nm, but cannot block an audible sound, whose average wavelength is about 100 cm. The bending of water waves around a post and the diffraction of light waves when passing through a slit in a screen are examples of diffraction patterns.

Refraction describes a change in the direction of a wave that occurs when the wave encounters a boundary between one medium and another provided that the media have either different wave velocities or indexes of refraction and provided that the wave arrives at some angle to the boundary other than perpendicular. At a sharp boundary, the change in direction is abrupt; however, if the transition from one medium to another is gradual, so that the velocity of the wave changes slowly, then the change in the wave's direction is also gradual. Therefore, a ray of light that passes obliquely from air to water changes its direction at the water's surface, but a ray that travels through air that has a temperature gradient will follow a bent path. A ray of light passing through a saturated solution of sugar (sucrose) and water, which has an index of refraction of 1.49, will not change direction appreciably on entering a colorless, transparent piece of quartz submersed in the solution because the quartz has an almost identical index of 1.51. The match in indexes makes the quartz nearly invisible in the sugar-water solution.

Another interesting phenomenon, the *Doppler effect,* accounts for the shift in the frequency of a wave when a wave source and an observer are in motion relative to each other compared with when they are at relative rest. This effect is most easily understood when the source is at rest in some medium and the observer is approaching the source at constant speed. The interval in time between each successive wave crest is shorter than it would be if the observer were at rest, and so the frequency observed is larger. The general rule, for observers moving at velocities much less than the velocity of the wave in its medium, is that the change in frequency depends only on the velocity of the observer relative to the source. Therefore, the shriek of an ambulance siren has a higher pitch when the source approaches and a lower pitch when the source recedes. For an observer following the ambulance at the same speed, the siren would sound normal. Similar shifts are observed for visible light.

Polarization is a property of light and of other transverse waves. *Transverse waves* are those in which the displacement of a test particle is always perpendicular to the direction in which the wave travels. When that displacement is always parallel to a particular direction, the wave is said to be *(linearly) polarized.* A ray of light emitted from a hot object, like a lamp filament or the sun, is unpolarized; such a ray consists of many component waves overlapped so that there is no special direction perpendicular to the ray in which a test particle is favored to move. The components of an unpolarized ray can be sorted to select such a special direction and so make one or more polarized rays. An unpolarized ray that is partly reflected and partly transmitted by an angled sheet of glass is split into rays that are polarized; an unpolarized ray can become polarized by going through a material that allows only waves corresponding to one special direction to pass through. Polarized sunglasses and stretched cellophane wrap are examples of polarizing materials.

STANDARD SET 5. Electric and Magnetic Phenomena

The electromagnetic force is one of only four fundamental forces; the others are the gravitational force and the forces that govern the strong and weak nuclear interactions. Electric and magnetic phenomena are well understood by scientists, and the unifying theory of the electromagnetic force is one of the great successes of science. The electromagnetic force accounts for the structure and for the unique chemical and physical properties of atoms and molecules. This force binds atoms and molecules and largely accounts for the properties of matter. Photons convey this force and electromagnetic energy.

Using electromagnetism for practical technological applications is taken for granted in modern society. Many devices of daily life, such as household appliances, computers, and equipment for communication, entertainment, and transportation, were developed from electromagnetic phenomena. Understanding the fundamental ideas of electricity and magnetism is basic to achieving success in a vast array of endeavors, from auto mechanics to nuclear physics.

Electricity and magnetism are now known to be two manifestations of a single phenomenon, the electromagnetic force. The originally separate theories explaining electricity and magnetism have been combined into a single theory of electromagnetism, whose predictive power is greater than that of either of the two previous, separate theories. The joining of these theories into a common mathematical framework is an example of how seemingly disparate phenomena can sometimes be unified in physics.

Studies of electric and magnetic phenomena build directly on the high school physics standards presented earlier and require a thorough understanding of the concepts of motion, forces, and conservation of energy. The subject of energy transport by waves is also important. Students in the lower grade levels are introduced to electricity as they learn that electric current can carry energy from one place to another. They also learn about light and the relationship between electricity and magnetism. To understand the concepts in Standard Set 5, students will need a strong grasp of beginning algebra and geometry. Basic trigonometry is also required for some of the advanced topics. Several topics covered in the lower grade levels may need to be reviewed as a part of teaching this standard set, particularly during the transition to standards-based education. In particular, the following facts are pertinent: (1) charge occurs in definite, discrete amounts; (2) charge comes in two varieties: positive and negative; and (3) the smallest amount of observable charge is the charge on an electron (or a proton).

Students should be acquainted with Newton's law of gravitation from standards studied previously (see Standard 1.e for grade two and Standard 4.c for grade five in Chapter 3 and standards 2.g and 4.e for grade eight in Chapter 4). Both Newton's law and Coulomb's law describe forces that diminish as the square of distance, and it may be helpful to compare those forces as a part of teaching some of the standards (see standards 1.e and 1.m* in this section). However, the comparison should be done with attention to the fundamental differences between the two types of forces, and certain points must be clearly understood to avoid sowing misconceptions. For example:

- Only the difference in electric or gravitational potential between two points has physical significance; the value of the potential at a particular point can be defined only relative to some reference point.

- The direction of an electric current is defined as the same as the direction of motion of charge carriers, conventionally assumed to be positive, although the charge carriers (the electrons) in wires are in fact negative. Therefore the direction of the electric current in wires is opposite to the direction of motion of the charge carriers.

- A *direct current* (DC) flows in one direction only, and an *alternating current* (AC) reverses at regular intervals.

- Ohm's law applies to conducting material under the assumption that resistance is independent of the magnitude and polarity of the potential difference (or of the applied electric field) across the material. The formulas used in Ohm's law to calculate an unknown amount of current, voltage, or resistance are $I = V/R$, $V = IR$, and $R = V/I$.

It may be helpful to describe *electric potential* as a measure of the tendency of a charged body to move from one point to another in an electrostatic field in the same way that *gravitational potential* is a measure of a body with mass to move from one point to another in a gravitational field. In both fields the work done to move the body does not depend on the path taken between the points but can be computed from the difference in the potential at the points.

As students solve simple circuit problems for this standard, they will also need to know the schematic representations of the various circuit elements, including a battery, a resistor, and a capacitor.

> **5. Electric and magnetic phenomena are related and have many practical applications. As a basis for understanding this concept:**
>
> **a.** *Students know* how to predict the voltage or current in simple direct current (DC) electric circuits constructed from batteries, wires, resistors, and capacitors.

Electric current I is the flow of net charge, and a complete, continuous path of current is called an *electric circuit*. If the charge carriers are positive, the electric current flows in the direction the carriers move; but if the carriers are negative, as they are in ordinary wires, the electric current flows in the opposite direction. Wires that carry currents are usually made of highly conducting metals, such as copper. If net charge q passes by a point a in a conducting wire in time t, the current I_a at that point is

$$I_a = q/t. \tag{eq. 32}$$

In the case of uniform current I, the rate of charge flow is the same through the entire length of the wire. Current is measured in units of amperes (A), which are equal to coulombs/second (A = C/s), the logical consequence of equation (32).

A particle with a charge q placed in an electric field will be subject to electrostatic forces and will have a potential energy. Moving the charge will change its potential energy from some value PE_a to PE_b, reflecting the work W_{ba} done by the electric field (see Standard 2.a in this section). Potential energy depends also on the magnitude of the charge being transported. A more convenient quantity is the potential energy per unit charge, which has a unique value at any point, independent of the actual charge of the particle in the electric field. This quantity is called *electric potential*, or just *potential*, and the difference V_{ab} between the potentials at two points a and b is the *voltage*. By this definition, voltage provides a measure of the work per unit charge required to move the charge between two points a and b in the field; alternatively, it represents the corresponding difference in potential energy per unit charge. This principle is expressed as

$$V_{ab} = V_a - V_b = W_{ba}/q = PE_a/q - PE_b/q. \tag{eq. 33}$$

Electric potential and voltage are measured in units of volt (V), which, as required by the preceding definition, is equal to joules per coulomb (J/C).

For a current-carrying wire, the potential difference between two points along the wire causes the current to flow in that segment.

> **5. b.** *Students know* how to solve problems involving Ohm's law.

Resistance, measured in ohms, of a conducting medium (conductor) is the opposition offered by the conductor to the flow of electric charge. A potential difference V is required to cause electrons to move continuously. Ohm's law gives the relationship between the current I that results when a voltage V is applied across a wire with resistance R. This law is expressed as

$$I = V/R. \tag{eq. 34}$$

Capacitors, which are devices for storing electrical charge, generally consist of two conductors with a potential difference that are separated by an insulator. A typical capacitor consists of two parallel metal plates insulated from each other by a *dielectric*, a material that does not conduct electricity. Capacitance C, the ability of a capacitor to store electric charge, can be measured in units of farads. The capacitance can be found from the following relation:

$$C = q/V, \qquad \text{(eq. 35)}$$

where q is the charge stored ($+q$ on one plate and $-q$ on the other) and V is the potential difference between the conducting surfaces. Based on equation (35), the unit of farad is defined as coulomb/volt (C/V).

> **5. c.** *Students know* any resistive element in a DC circuit dissipates energy, which heats the resistor. Students can calculate the power (rate of energy dissipation) in any resistive circuit element by using the formula Power = *IR* (potential difference) × *I* (current) = *I²R*.

Electric power P is defined as the rate of dissipation of electric energy, or the rate of production of heat energy, in a resistor and is given by Joule's law, in which

$$P = IV. \qquad \text{(eq. 36)}$$

Through the use of Ohm's law, this equation can also be written as $P = I^2 R$ or $P = V^2/R$. Power is measured in watts, where 1 watt = 1 ampere-volt (W = A-V) = 1 joule/second.

Dissipation of energy as heat is a consequence of electrical resistance. In other words *electric power* is equivalent to the work per second that must be done to maintain an electric current. Alternatively, *power* is the rate at which electrical energy is transferred from the source to other parts of the circuit. The unit of kilowatt hour (kWH) is sometimes used commercially to represent energy production and consumption, where 1 kWH = 3.6×10^6 J.

> **5. d.** *Students know* the properties of transistors and the role of transistors in electric circuits.

Semiconductors are materials with an energy barrier such that only electrons with energy above a certain amount can "flow." As the temperature rises, more electrons are free to move through these materials. A transistor is made of a combination of differently "doped" materials arranged in a special way. Transistors can be used to control large current output with a small bias voltage. A common role of transistors in electric circuits is that of amplifiers. In that role transistors have almost entirely replaced vacuum tubes that were widely used in early radios, television sets, and computers.

> **5. e.** *Students know* charged particles are sources of electric fields and are subject to the forces of the electric fields from other charges.

Electrostatic force represents an interaction across space between two charged bodies. The magnitude of the force is expressed by a relationship similar to that for the gravitational force between two bodies with mass. For both gravity and electricity, the force varies inversely as the square of the distance between the two bodies. For two charges q_1 and q_2 separated by a distance r, the relationship is called Coulomb's law,

$$F = kq_1q_2/r^2 , \qquad \text{(eq. 37)}$$

where k is a constant. Customary units for charge are coulombs (C), in which case $k = 9 \times 10^9 \text{Nm}^2/\text{C}^2$.

An electric field is a condition produced in space by the presence of charges. A field is said to exist in a region of space if a force can be measured on a test charge in the region. Many different and complicated distributions of electric charge can produce the same simple motion of a test charge and therefore the same simple field; for that reason it is usually easier to study first the effect of a model field on a test charge and to consider only later what distribution of other charges might produce that field.

> **5. f.** *Students know* magnetic materials and electric currents (moving electric charges) are sources of magnetic fields and are subject to forces arising from the magnetic fields of other sources.

A magnetic force exists between magnets or current-carrying conductors or both. A stationary charge does not produce magnetic forces. Furthermore, no evidence for the existence of magnetic monopoles, which would be the magnetic equivalent of electric charges, has yet been found. Iron and other materials that can be magnetized have domains in which the combined motion of electrons produces the equivalent of small magnets in the metal. When many of these domains are aligned, the entire metal object becomes a strong magnet. Therefore, to the best of scientific knowledge, all magnetic effects result from the motion of electrical charges.

The concept of a field applies to magnetism just as it does to electricity (see Standard 5.e in this section). Magnetic fields are generated either by magnetic materials or by electric currents caused by the motion of charged particles. A standard unit for the magnetic field strength is the Tesla (T). Electric charges moving through a magnetic field experience a magnetic force. The direction of the magnetic force is always perpendicular to the line of motion of the electric charges. The force is at maximum when the direction of motion of the electric charges (their velocity vector) is perpendicular to the magnetic field and at zero when the two are parallel.

5. g. *Students know* how to determine the direction of a magnetic field produced by a current flowing in a straight wire or in a coil.

The direction of a magnetic field is by convention taken to be outward from a north pole and inward from a south pole. The right-hand rule finds the direction of the magnetic field produced by a current flowing in a wire or coil. To find the direction in a wire, a student wraps the fingers of the right hand around the wire with the thumb pointing in the direction in which the electric current flows (in a wire electrons and electric current move in opposite directions). The fingers encircling the wire then point in the direction of the magnetic field outside the wire. The same rule will find the direction of the magnetic field inside a coil if one imagines that the right hand wraps around a wire that forms one of the loops that make up the coil. A different rule using the right hand also works for coils. The coil is held in the palm of the right hand with the fingers wrapped around the coil and pointing in the direction in which the electric current flows through the loops. The thumb then points in the direction of the magnetic field inside the coil.

5. h. *Students know* changing magnetic fields produce electric fields, thereby inducing currents in nearby conductors.

The concept of electromagnetic induction is based on the observation that changing magnetic fields create electric fields, just as changing electric fields are sources of magnetic fields. In a conductor these induced electric fields can drive a current. The direction of the induced current is always such as to oppose the changing magnetic field that caused it. This principle is called Lenz's law.

5. i. *Students know* plasmas, the fourth state of matter, contain ions or free electrons or both and conduct electricity.

A *plasma* is a mixture of positive ions and free electrons that is electrically neutral on the whole but that can conduct electricity. A plasma can be created by very high temperatures when molecules disassociate and their constituent atoms further break up into positively charged ions and negatively charged electrons. Much of the matter in the universe is in stars in the form of plasma, a mixture of electrified fragments of atoms. Plasma is considered a fourth state of matter, as fundamental as solid, liquid, and gas.

5. j.* *Students know* electric and magnetic fields contain energy and act as vector force fields.

Both the electric field **E** and the magnetic field **B** are vector fields; therefore, they have a magnitude and a direction. The fields from matter whose distributions in space and in velocity do not change with time are easy to visualize; for example, charges fixed in space, steady electric currents in wires, or permanent magnets. Electric fields from matter like this are generally represented by "lines of force" that start on positive charges and end on negative charges but never form closed loops (see Standard 5.m*, which appears later in this section). In contrast, the lines for magnetic fields always form closed loops; they never start and end—magnetic field lines do not have terminal points. Even the magnetic field lines around simple bar magnets, which are typically drawn as emanating from the north pole and entering the south pole, in fact continue through the body of the magnet to form closed loops.

The reason magnetic fields form loops while electric fields do not has to do with their different sources in matter at rest. Electric fields come from point charges, and magnetic fields come from point dipoles, which are more complicated; no sources of magnetic field with the simple properties of charge—that is, no magnetic monopoles—are known to exist. The direction in which an electric field points along a line of force is away from positive charge and toward negative charge; the direction in which a magnetic field (that is due to a current) points along a closed loop can be found by the right-hand rule (see Standard 5.g, which appears earlier in this section).

Electric and magnetic fields are associated with the existence of potential energy. The fields are usually said to *contain* energy. For example, the potential energy of a system of two charges q_1 and q_2 located a distance r apart, is given by

$$PE = kq_1q_2/r. \tag{eq. 38}$$

In general, the potential energy of a system of fixed-point charges is defined as the work required to assemble the system bringing each charge in from an infinite distance.

> **5. k.*** *Students know* the force on a charged particle in an electric field is q**E**, where **E** is the electric field at the position of the particle and q is the charge of the particle.

The electric field strength **E** at a given point is defined as the force experienced by a unit positive charge, $\mathbf{E} = \mathbf{F}/q$. The units of **E** are newton/coulomb (N/C). By this definition the force experienced by a charged particle is

$$\mathbf{F} = q\mathbf{E}, \tag{eq. 39}$$

where q is the magnitude of the particle's charge in coulombs and **E** is the electric field at the position of the charged particle.

> **5. l.*** *Students know* how to calculate the electric field resulting from a point charge.

Coulomb's law is used in calculating the electric field caused by a point charge. According to equation (39), $\mathbf{E} = \mathbf{F}/q$, the magnitude of the field produced by a point charge q_1 is found by substituting equation (37) for **F** and dividing by the magnitude of the positive test charge q_2, which gives

$$E = kq_1/r^2. \tag{eq. 40}$$

The direction of **E** is determined by the type of the source charge q_1, so that the vector is away from the positive charge (+) and toward the negative charge (−). (Remember that by definition the *field strength* is the force per unit of positive test charge.)

> **5. m.*** *Students know* static electric fields have as their source some arrangement of electric charges.

The existence of a static electric field in a region of space implies a distribution of charges as the source. Conversely, any set of charges or charged surfaces sets up an electric field in the space around the charge. The customary first step in visualizing an electric field is to draw smooth curves, each of which contains only points of equal electric potential. Electric field lines ("lines of force") can then be drawn as curves that are everywhere perpendicular to the curves of equal potential. Electric field lines are assigned a direction that runs from regions of high potential to low and, therefore, from positive point charges to negative ones. The lines of force represent the path a particle with a small positive charge would take if released in the field.

The method used in deriving equation (40) can be used, in principle, to determine the field produced from any distribution of charges. At each point a net vector **E** is obtained by summing the vector contributions from each charge. This process can be readily done for a two-charge system in which the geometry is relatively simple. For more complicated distributions the methods of calculus are generally required to obtain the field.

> **5. n.*** *Students know* the magnitude of the force on a moving particle (with charge q) in a magnetic field is qvB sin(a), where a is the angle between **v** and **B** (v and B are the magnitudes of vectors **v** and **B**, respectively), and students use the right-hand rule to find the direction of this force.

The force on a moving particle of charge q traveling at velocity v in a magnetic field B is given by

$$F = qvB \sin(a) ,\qquad\text{(eq. 41)}$$

where a is the angle between the direction of the motion of the charged particle and the direction of the magnetic field. (If $a = 0$, then the particle is traveling parallel to the direction of the field and the magnetic force on it is zero.) The maximum force is obtained when the particle is traveling perpendicular to the magnetic field. Students can determine the direction of the magnetic force through the use of the right-hand rule. The magnetic force is perpendicular to both the direction of motion of the charge and to the direction of the magnetic field. Equation (41) shows that Tesla, a standard unit for the magnetic field mentioned previously, is equal to 1 N-s/C-m (see the discussion for Standard 5.f, which appears previously in this section).

5. o.* *Students know* how to apply the concepts of electrical and gravitational potential energy to solve problems involving conservation of energy.

In standards 2.a and 2.b in this section, students learned that if a stone is raised from Earth's surface, the work done against Earth's gravitational attraction is stored as potential energy in the system of stone plus Earth. If the stone is released, the stored potential energy is transformed into kinetic energy, which steadily increases as the stone moves faster toward Earth. Once the stone comes to rest, this kinetic energy will ultimately be transformed into thermal energy. A similar situation exists in electrostatics. If the separation between two opposite charges is increased, work must be performed. The work is positive if the charges are opposite and negative. The energy represented by this work can be thought of as stored in the system of charges as electric potential energy (see also Standard 5.j* in this section) and, like gravitational potential energy, may be transformed into other forms, such as kinetic and thermal energy.

A simple example is a charge q moving freely between point a and point b, with a potential difference V_{ab} between the two points. If q is positive, the change in electric potential energy can be found from equation (35) and is

$$\Delta PE = qV_{ab}. \qquad \text{(eq. 42)}$$

By conservation of energy a corresponding amount of the kinetic energy is acquired, or released, by the charge at point b such that

$$\Delta KE = \Delta PE = qV_{ab}. \qquad \text{(eq. 43)}$$

Through substitution of the standard expression $\frac{1}{2}mv^2$ for the kinetic energy, a variety of predictions can be made, assuming the accelerating potential does not result in velocities approaching the velocity of light. The final velocity v can be found if the charge q, the mass m, and the potential V are known. This method of imparting energy to charged particles is applied in such devices as television sets and in accelerators used in modern atomic and nuclear experiments.

Notes

1. *Science Content Standards for California Public Schools, Kindergarten Through Grade Twelve.* Sacramento: California Department of Education, 2000.
2. *Science Safety Handbook for California Public Schools.* Sacramento: California Department of Education, 1999.

Chemistry

A sign in a university professor's office asks, What in the world isn't chemistry? Although meant to amuse, this question has a world of truth behind it. High school students come into contact with chemistry every day, often without realizing it. Discussions of daily interactions with chemistry often provide an entry into teaching the subject in high school. Although relating chemistry to daily life is helpful, this approach does not diminish the need for students to have a high level of readiness before entering the class. Of paramount importance is a firm grounding in algebra.

Chemistry is a sequential, hierarchical science that is descriptive and theoretical. It requires knowing the macroscopic properties of matter and the microscopic properties of matter's constituent particles. Although chemical demonstrations may engage students, going beyond a superficial appreciation of chemistry is a critical step. Chemistry requires high-level problem-solving skills, such as designing experiments and solving word problems. For students to learn concepts of chemistry, they must learn new vocabulary, including the rules for naming simple compounds and ions.

Students can discover chemistry's tremendous power to explain the nature of matter and its transformations when they study the periodic table of the elements. Students who move beyond a trivial treatment of the discipline can explore the many useful, elegant, and even beautiful aspects of chemistry. Bringing students to this understanding is a great achievement.

STANDARD SET 1. Atomic and Molecular Structure

Starting with atomic and molecular structure in the study of high school chemistry is important because this topic is a foundation of the discipline. However, because structural concepts are highly theoretical and deal with the quantum realm, they can be hard to relate to real-world experience. Ideally, from grades three through eight, students have been gradually introduced to the atomic theory; and by the end of the eighth grade, they should have covered the major concepts in the structure of atoms and molecules. By the time students reach high school, they should be familiar with basic aspects of this theory.

The study of structure can begin with the simplest element, hydrogen. Students can progress from a simple model of the atom (see standards 1.a, 1.d, 1.e, 1.h*, and 11.a in this section) to the historic Bohr model (see Standard 1.i* in this section) and, finally, to a quantum mechanical model (see standards 1.g* and 1.j* in this section), which is the picture of the atom that students should ultimately develop. Students should learn that the quantum mechanical model takes into account the particle and wave properties of the electron and uses mathematical equations to solve for electron energies and regions of electron density.

Students should understand that the energy carried by electrons either within an atom or as electricity can be transformed into light energy. Those who have completed high school physics will be familiar with the properties of electromagnetic waves. In chemistry students learn to apply the equations $E = h\nu$ and $c = \lambda\nu$. Students without significant training in physics will need to understand electromagnetic radiation as energy, frequency, and wavelength. The necessary mathematical background for the study of chemistry includes algebraic isolation of variables, use of conversion factors, and manipulation of exponents, all of which are covered in the mathematics standards for the middle grades.

> **1. The periodic table displays the elements in increasing atomic number and shows how periodicity of the physical and chemical properties of the elements relates to atomic structure. As a basis for understanding this concept:**
>
> **a.** *Students know* how to relate the position of an element in the periodic table to its atomic number and atomic mass.

An atom consists of a nucleus made of protons and neutrons that is orbited by electrons. The number of protons, not electrons or neutrons, determines the unique properties of an element. This number of protons is called the element's atomic number. Elements are arranged on the periodic table in order of increasing atomic number. Historically, elements were ordered by atomic mass, but now scientists know that this order would lead to misplaced elements (e.g., tellurium and iodine) because differences in the number of neutrons for isotopes of the same element affect the atomic mass but do not change the identity of the element.

> **1. b.** *Students know* how to use the periodic table to identify metals, semimetals, nonmetals, and halogens.

Most periodic tables have a heavy stepped line running from boron to astatine. Elements to the immediate right and left of this line, excluding the metal aluminum, are semimetals and have properties that are intermediate between metals and nonmetals. Elements further to the left are metals. Those further to the right are nonmetals. Halogens, which are a well-known family of nonmetals, are found in Group 17 (formerly referred to as Group VIIA). A group, also sometimes called a "family," is found in a vertical column in the periodic table.

> **1. c.** *Students know* how to use the periodic table to identify alkali metals, alkaline earth metals and transition metals, trends in ionization energy, electronegativity, and the relative sizes of ions and atoms.

A few other groups are given family names. These include the alkali metals (Group 1), such as sodium and potassium, which are soft and white and extremely reactive chemically. Alkaline earth metals (Group 2), such as magnesium and calcium, are found in the second column of the periodic table. The transition metals (Groups 3 through 12) are represented by some of the most common metals, such as iron, copper, gold, mercury, silver, and zinc. All these elements have electrons in their outer *d* orbitals.

Electronegativity is a measure of the ability of an atom of an element to attract electrons toward itself in a chemical bond. The values of electronegativity calculated for various elements range from one or less for the alkali metals to three and one-half for oxygen to about four for fluorine. *Ionization energy* is the energy it takes to remove an electron from an atom. An element often has multiple ionization energies, which correspond to the energy needed to remove first, second, third, and so forth electrons from the atom. Generally in the periodic table, ionization energy and electronegativity increase from left to right because of increasing numbers of protons and decrease from top to bottom owing to an increasing distance between electrons and the nucleus. Atomic and ionic sizes generally decrease from left to right and increase from top to bottom for the same reasons. Exceptions to these general trends in properties occur because of filled and half-filled subshells of electrons.

> **1. d.** *Students know* how to use the periodic table to determine the number of electrons available for bonding.

Only electrons in the outermost energy levels of the atom are available for bonding; this outermost bundle of energy levels is often referred to as the *valence shell* or *valence shell of orbitals.* All the elements in a group have the same number of electrons in their outermost energy level. Therefore, alkali metals (Group 1) have one electron available for bonding, alkaline earth metals (Group 2) have two, and elements in Group 13 (once called Group III) have three. Unfilled energy levels are also available for bonding. For example, Group 16, the chalcogens, has room for two more electrons; and Group 17, the halogens, has room for one more electron to fill its outermost energy level.

To find the number of electrons available for bonding or the number of unfilled electron positions for a given element, students can examine the combining ratios of the element's compounds. For instance, one atom of an element from Group 2 will most often combine with two atoms of an element from Group 17 (e.g., $MgCl_2$) because Group 2 elements have two electrons available for bonding, and Group 17 elements have only one electron position open in the outermost energy level. (Note that some periodic tables indicate an element's electron configuration or preferred oxidation states. This information is useful in determining how many electrons are involved in bonding.)

> **1. e.** *Students know* the nucleus of the atom is much smaller than the atom yet contains most of its mass.

The volume of the hydrogen nucleus is about one trillion times less than the volume of the hydrogen atom, yet the nucleus contains almost all the mass in the form of one proton. The diameter of an atom of any one of the elements is about 10,000 to 100,000 times greater than the diameter of the nucleus. The mass of the atom is densely packed in the nucleus.

The electrons occupy a large region of space centered around a tiny nucleus, and so it is this region that defines the volume of the atom. If the nucleus (proton) of a hydrogen atom were as large as the width of a human thumb, the electron would be on the average about one kilometer away in a great expanse of empty space. The electron is almost 2,000 times lighter than the proton; therefore, the large region of space occupied by the electron contains less than 0.1 percent of the mass of the atom.

> **1. f.*** *Students know* how to use the periodic table to identify the lanthanide, actinide, and transactinide elements and know that the transuranium elements were synthesized and identified in laboratory experiments through the use of nuclear accelerators.

The lanthanide series, or rare earths, and the actinide series, all of which are radioactive, are separated for reasons of practical display from the main body of the periodic table. If these two series were inserted into the main body, the table would be wider by 14 elements and less manageable for viewing. Within each of these series, properties are similar because the configurations of outer electrons are similar. As a general rule elements in both series appear to have three electrons available for bonding. They combine with halogens to form compounds with the general formula MX_3, such as $LaFz_3$. The transactinide elements begin with rutherfordium, element 104.

All the elements with atomic numbers greater than 92 were first synthesized and identified in experiments. These experiments required the invention and use of accelerators, which are electromagnetic devices designed to create new elements by accelerating and colliding the positively charged nuclei of atoms. Ernest O. Lawrence, at the University of California, Berkeley, invented one of the most useful nuclear accelerators, the cyclotron. Many transuranium elements, such as 97-berkelium, 98-californium, 103-lawrencium, and 106-seaborgium, were first created and identified at the adjacent Lawrence Berkeley National Laboratory. Today a few transuranium elements are produced in nuclear reactors. For example, hundreds of metric tons of plutonium have been produced in commercial nuclear reactors.

> **1. g.*** *Students know* how to relate the position of an element in the periodic table to its quantum electron configuration and to its reactivity with other elements in the table.

Each element has a unique electron configuration (also known as quantum electron configuration) that determines the properties of the element. Quantum mechanical calculations predict these electron energy states, which provide the theoretical justification for the organization of the periodic table, previously organized on the basis of chemical properties.

Students can learn the principal quantum numbers—which are 1, 2, 3, 4, 5, 6, and 7—and the corresponding periods, or horizontal rows, on the periodic table. They can learn the angular momentum quantum numbers that give rise to s, p, d, and f subshells of orbitals and the rules for the sequence of orbital filling. The electrons in the highest energy orbitals with the same principal quantum number are the *valence electrons.* For example, for all elements in Group 1, the valence electron configuration is ns^1, where n is the principal quantum number. Analogously, all elements in Group 16 have valence electron configurations of ns^2np^4. Particular configurations of valence electrons are associated with regular patterns in chemical reactivity. Generally, those elements with one electron in excess or one electron short of a full octet in the highest occupied energy level, the alkali metals and halogens, respectively, are the most reactive.

> **1. h.*** *Students know* the experimental basis for Thomson's discovery of the electron, Rutherford's nuclear atom, Millikan's oil drop experiment, and Einstein's explanation of the photoelectric effect.

In 1887 J. J. Thomson performed experiments from which he concluded that cathode rays are streams of negative, identical particles, which he named *electrons.* In 1913 E. R. Rutherford headed a group that shot a beam of helium nuclei, or alpha particles, through a very thin piece of gold foil; the unexpectedly large deflections of some helium nuclei led to the hypothesis that the charged mass of each gold atom is concentrated in a small central nucleus. Robert Millikan confirmed Thomson's conclusion that electrons are identical particles when he balanced tiny, electrically charged oil droplets between electric and gravitational fields and so discovered that the droplets always contained charge equal to an integral multiple of a single unit.

Albert Einstein found he could explain the photoelectric effect, in which light ejects electrons from metal surfaces, by proposing that light consists of discrete bundles of energy, or *photons,* each photon with an energy directly proportional to the frequency of the light, and by proposing that each photon could eject one and only one electron. Photons with sufficient energy will eject an electron whose kinetic energy is equal to the photon energy minus the energy required to free the electron from the metal. If the frequency of the light and therefore the energy of each proton is too low to free an electron, then merely increasing the light's intensity (that is, merely producing more photons) does not cause electrons to eject.

> **1. i.*** *Students know* the experimental basis for the development of the quantum theory of atomic structure and the historical importance of the Bohr model of the atom.

Niels Bohr combined the concepts of Rutherford's nuclear atom and Einstein's photons with several other ideas to develop a model that successfully explains the observed spectrum, or wavelengths, of electromagnetic radiation that is emitted when a hydrogen atom falls from a high energy state to a low energy state. In classical physics all accelerating charges emit energy. If electrons in an atom behaved in this way, light of ever-decreasing frequencies would be emitted from atoms. Bohr explained why this phenomenon does not occur when he suggested that electrons in atoms gain or lose energy only by making transitions from discrete energy levels. This idea was a key to the development of nonclassical descriptions of atoms. Louis de Broglie advanced the understanding of the nature of matter by proposing that particles have wave properties. On the basis of these ideas, Erwin Schrödinger and others developed quantum mechanics, a theory that describes and predicts atomic and nuclear phenomena.

> **1. j.*** *Students know* that spectral lines are the result of transitions of electrons between energy levels and that these lines correspond to photons with a frequency related to the energy spacing between levels by using Planck's relationship $(E = h\nu)$.

The Bohr model gives a simple explanation of the spectrum of the hydrogen atom. An electron that loses energy in going from a higher energy level to a lower one emits a photon of light, with energy equal to the difference between the two energy levels. Transitions of electrons from higher energy states to lower energy states yield *emission*, or *bright-line, spectra. Absorption spectra* occur when electrons jump to higher energy levels as a result of absorbing photons of light. When atoms or molecules absorb or emit light, the absolute value of the energy change is equal to hc/λ, where h is a number called *Planck's constant*, c is the speed of light, and λ is the wavelength of light emitted, yielding Planck's relationship $E = h\nu$.

The transition from the Bohr model to the quantum mechanical model of the atom requires students to be aware of the probabilistic nature of the distribution of electrons around the atom. A "dart board" made of concentric rings can serve as a two-dimensional model of the three-dimensional atom. A graph as a function of radius of the number of random hits by a dart can be compared with a similar graph as a function of radius of the probability density of the electron in a hydrogen atom.

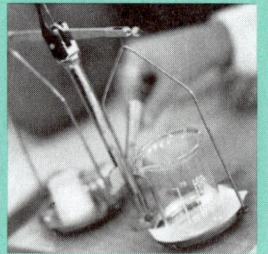

STANDARD SET 2. Chemical Bonds

Standard Set 2 deals with two distinct topics: chemical bonds and intermolecular attractive forces, such as hydrogen bonds. A logical place to begin the study of this standard is with a discussion of the chemical bond. A key point to emphasize is that when atoms of two different elements join to form a covalent bond, energy is almost always released. Conversely, breaking bonds always requires the addition of energy. Students should understand that the sum of these two processes determines the net energy released or absorbed in a chemical reaction.

This standard set requires a basic knowledge of electrostatics and electronegativity and a thorough knowledge of the periodic table. After studying standards for chemistry for the elementary grades, students should know that matter is made of atoms and that atoms combine to form molecules. Students can also be expected to know that atoms consist of protons, neutrons, and electrons. Although knowledge of complex mathematics is not required for this standard, some background in three-dimensional geometry will be helpful.

> **2. Biological, chemical, and physical properties of matter result from the ability of atoms to form bonds from electrostatic forces between electrons and protons and between atoms and molecules. As a basis for understanding this concept:**
>
> **a.** *Students know* atoms combine to form molecules by sharing electrons to form covalent or metallic bonds or by exchanging electrons to form ionic bonds.

In the localized electron model, a covalent bond appears as a shared pair of electrons contained in a region of overlap between two atomic orbitals. Atoms (usually nonmetals) of similar electronegativities can form covalent bonds to become molecules. In a covalent bond, therefore, bonding electron pairs are localized in the region between the bonded atoms. In metals valence electrons are not localized to individual atoms but are free to move to temporarily occupy vacant orbitals on adjacent metal atoms. For this reason metals conduct electricity well.

When an electron from an atom with low electronegativity (e.g., a metal) is removed by another atom with high electronegativity (e.g., a nonmetal), the two atoms become oppositely charged ions that attract each other, resulting in an ionic bond. Chemical bonds between atoms can be almost entirely covalent, almost entirely ionic, or in between these two extremes. The triple bond in nitrogen molecules (N_2) is nearly 100 percent covalent. A salt such as sodium chloride (NaCl) has bonds that are nearly completely ionic. However, the electrons in gaseous hydrogen chloride are shared somewhat unevenly between the two atoms. This kind of bond is called *polar covalent*.

(Note that elements in groups 1, 2, 16, and 17 in the periodic table usually gain or lose electrons through the formation of either ionic or covalent bonds, resulting in eight outer shell electrons. This behavior is sometimes described as "the octet rule.")

> **2. b.** *Students know* chemical bonds between atoms in molecules such as H_2, CH_4, NH_3, H_2CCH_2, N_2, Cl_2, and many large biological molecules are covalent.

Organic and biological molecules consist primarily of carbon, oxygen, hydrogen, and nitrogen. These elements share valence electrons to form bonds so that the outer electron energy levels of each atom are filled and have electron configurations like those of the nearest noble gas element. (Noble gases, or inert gases, are in the last column on the right of the periodic table.) For example, nitrogen has one lone pair and three unpaired electrons and therefore can form covalent bonds with three hydrogen atoms to make four electron pairs around the nitrogen. Carbon has four unpaired electrons and combines with hydrogen, nitrogen, and oxygen to form covalent bonds sharing electron pairs.

The great variety of combinations of carbon, nitrogen, oxygen, and hydrogen make it possible, through covalent bond formation, to have many compounds from just these few elements. Teachers can use ball and stick or gumdrop and toothpick models to explore possible bonding combinations.

> **2. c.** *Students know* salt crystals, such as NaCl, are repeating patterns of positive and negative ions held together by electrostatic attraction.

The energy that holds ionic compounds together, called *lattice energy,* is caused by the electrostatic attraction of *cations,* which are positive ions, with *anions,* which are negative ions. To minimize their energy state, the ions form repeating patterns that reduce the distance between positive and negative ions and maximize the distance between ions of like charges.

> **2. d.** *Students know* the atoms and molecules in liquids move in a random pattern relative to one another because the intermolecular forces are too weak to hold the atoms or molecules in a solid form.

In any substance at any temperature, the forces holding the material together are opposed by the internal energy of particle motion, which tends to break the substance apart. In a solid, internal agitation is insufficient to overcome intermolecular or interatomic forces. When enough energy is added to the solid, the kinetic energy of the atoms and molecules increases sufficiently to overcome the attractive forces between the particles, and they break free of their fixed lattice positions. This change, called *melting,* forms a liquid, which is disordered and nonrigid. The particles in the liquid are free to move about randomly although they remain in contact with each other.

> **2. e.** *Students know* how to draw Lewis dot structures.

A Lewis dot structure shows how valence electrons and covalent bonds are arranged between atoms in a molecule. Teachers should follow the rules for drawing Lewis dot diagrams provided in a chemistry textbook. Students should be able to use the periodic table to determine the number of valence electrons for each element in Groups 1 through 3 and 13 through 18. Carbon, for example, would have four valence electrons. Lewis dot diagrams represent each electron as a dot or an *x* placed around the symbol for carbon, which is C. A covalent bond is shown as a pair of dots, or *x*'s, representing a pair of electrons. For example, a Lewis dot diagram for methane, which is CH_4, would appear as shown in Figure 3.

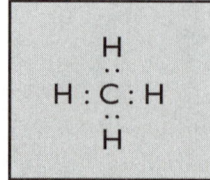

Fig. 3. Lewis Dot Diagram

Lewis dot diagrams provide a method for predicting correct combining ratios between atoms and for determining aspects of chemical bonds, such as whether they are covalent or consist of single, double, or triple bonds.

> **2. f.*** *Students know* how to predict the shape of simple molecules and their polarity from Lewis dot structures.

Using information obtained from Lewis dot structures of covalently bonded molecules, students can predict the overall geometry of those molecules. This model assumes that valence electron pairs repel each other and that atoms covalently bonded around a central atom position themselves as far apart as possible while maintaining the covalent bond. The model also assumes that double or triple bonds define a single electronic region.

To predict shapes, students start with a correct Lewis dot structure. From the number of electron pairs or regions, both bonded and nonbonded, students can determine the molecular geometry of the molecule because Lewis dot structures, although drawn in two dimensions, represent the three-dimensional symmetry of the molecule. A symmetrical distribution of the electron clouds around a central atom leads to a nonpolar molecule in which charge is evenly distributed. Students should be able to predict that a molecule such as methane, with one carbon and four hydrogen atoms, forms a tetrahedral shape and because of its symmetry is a nonpolar molecule.

> **2. g.*** *Students know* how electronegativity and ionization energy relate to bond formation.

During bond formation atoms with large electronegativity values, such as fluorine and oxygen, attract electrons away from lower electronegativity atoms, such as the alkali metals. The difference in electronegativity between the two bonding atoms gives information on how evenly an electron pair is shared. A large difference in electronegativity leads to an ionic bond, with essentially no sharing of electrons. This phenomenon usually occurs between metal and nonmetal atoms. A small difference in electronegativity leads to a covalent bond with more equal sharing of electrons. This result typically occurs between two nonmetal atoms. Electronegativity is related to *ionization energy,* the energy needed to remove an electron from an isolated gaseous atom, leaving a positively charged ion. High ionization energies usually correlate with large electronegativities.

> **2. h.*** *Students know* how to identify solids and liquids held together by van der Waals forces or hydrogen bonding and relate these forces to volatility and boiling/melting point temperatures.

Liquids and solids that are held together not by covalent or ionic bonds but only by weaker intermolecular forces tend to have low to moderate melting and boiling points and to be from somewhat to very volatile. The *volatility* of a substance means how readily it evaporates at ordinary temperatures and pressures.

Two kinds of intermolecular forces are hydrogen bonding and van der Waals attractions (often referred to as London dispersion forces). Hydrogen bonding is essential to life and gives water many of its unusual properties. A hydrogen bond occurs when a hydrogen atom on one molecule, bonded directly to a highly electronegative atom (fluorine, oxygen, or nitrogen), is weakly attracted to the electronegative atom on a neighboring molecule. In the important case of water, this attraction exists between the hydrogen on one water molecule and the oxygen on a neighboring water molecule. This attraction happens because of water's polar nature and bent shape.

Van der Waals forces exist between all molecules, polar or nonpolar. Even when the molecule is nonpolar, the electrons move around and may sometimes find themselves temporarily closer to one nucleus than to the other. The atom with the greater share of the electron density becomes, for an instant, slightly negatively charged, and the other atom becomes a little bit positive. If the same thing happens in a nearby molecule, the positive and negative centers on the two molecules temporarily attract each other. Bigger molecules have more electrons, and their van der Waals forces are greater. This phenomenon leads to molecules with lower volatility and higher melting and boiling temperatures.

STANDARD SET 3. Conservation of Matter and Stoichiometry

Standard Set 3 demands more facility with mathematics than do the previous two chemistry standard sets. For this reason students need prerequisite mathematical skills in two broad categories. First, they must be able to manipulate very large and very small numbers by using exponents as expressed in scientific notation, and they should learn the rules of significant digits when reporting measurements and the results of calculations. Second, students must be able to manipulate simple equations in symbolic form, such as the isolation of variables, and to write equations numerically with correct units. Handling units successfully is necessary for problems involving mole-to-mass and mass-to-mole conversions. The ability to make other types of unit conversion and to square and cube linear measurements will also be required. An understanding of ratios will help students to see the logic behind problems related to Standard Set 3.

Simple metric conversions are relatively easy to grasp, and they contain the basic elements of stoichiometric calculations. For example, to convert 510 nanometers to meters, students write the unknown unit, which in this case is meters; set it equal to the given unit, which in this case is nanometers; and then multiply by the correct conversion factor, making sure that the desired unit is in the numerator and the unit to be cancelled is in the denominator. The following equation demonstrates this procedure:

$$510 \text{ nm} = 510 \text{ nm} \times (10^{-9} \text{ m}/1 \text{ nm}) = 510 \times 10^{-9} \text{ m} = 5.1 \times 10^{-7} \text{ m}$$

This technique converts any units however complicated. For example, centimeters can be converted to nanometers by multiplying by two factors: one that converts from centimeters to meters and another that converts from meters to nanometers. Converting to an area requires the square of a conversion factor; converting to a volume requires the cube. Once students have tackled simple problems, they can move on to more complex stoichiometric relations. They will also need to learn to balance equations easily.

> **3. The conservation of atoms in chemical reactions leads to the principle of conservation of matter and the ability to calculate the mass of products and reactants. As a basis for understanding this concept:**
>
> **a.** *Students know* how to describe chemical reactions by writing balanced equations.

Reactions are described by balanced equations because all the atoms of the reactants must be accounted for in the reaction products. An equation with all correct chemical formulas can be balanced by a number of methods, the simplest being by inspection. Given an unbalanced equation, students can do an inventory to determine how many of each atom are on each side of the equation. If the result is not equal for all atoms, coefficients (not subscripts) are changed until balance is achieved. Sometimes, reactions refer to substances with written names rather than to chemical symbols. Students should learn the rules of chemical nomenclature. This knowledge can be acquired in stages as new categories of functional groups are introduced.

> **3. b.** *Students know* the quantity *one mole* is set by defining one mole of carbon-12 atoms to have a mass of exactly 12 grams.

The mole concept is often difficult for students to understand at first, but they can be taught that the concept is convenient in chemistry just as a dozen is a convenient concept, or measurement unit, in the grocery store. The mole is a number. Specifically, a *mole* is defined as the number of atoms in 12 grams of carbon-12. When atomic masses were assigned to elements, the mass of 12 grams of carbon-12 was selected as a standard reference to which the masses of all other elements are compared. The number of atoms in 12 grams of carbon-12 is defined as one *mole*, or conversely, if one mole of ^{12}C atoms were weighed, it would weigh exactly 12 grams. (Note that carbon, as found in nature, is a mixture of isotopes, including atoms of carbon-12, carbon-13, and trace amounts of carbon-14.) The definition of the mole refers to pure carbon-12.

The *atomic mass* of an element is the weighted average of the mass of one mole of its atoms based on the abundance of all its naturally occurring isotopes. The atomic mass of carbon is 12.011 grams. If naturally occurring carbon is combined with oxygen to form carbon dioxide, the mass of one mole of naturally occurring oxygen can be determined from the combining mass ratios of the two elements. For example, the weight, or atomic mass, of one mole of oxygen containing mostly oxygen-16 and a small amount of oxygen-18 is 15.999 grams.

> **3. c.** *Students know* one mole equals 6.02×10^{23} particles (atoms or molecules).

A mole is a very large number. Standard 3.b describes the mole as the number of atoms in 12 grams of ^{12}C. The number of atoms in a mole has been found experimentally to be about 6.02×10^{23}. This number, called Avogadro's number, is known to a high degree of accuracy.

> **3. d.** *Students know* how to determine the molar mass of a molecule from its chemical formula and a table of atomic masses and how to convert the mass of a molecular substance to moles, number of particles, or volume of gas at standard temperature and pressure.

The molar mass of a compound, which is also called either the *molecular mass* or *molecular weight,* is the sum of the atomic masses of the constituent atoms of each element in the molecule. Molar mass is expressed in units of grams per mole. The periodic table is a useful reference for finding the atomic masses of each element. For example, one mole of carbon dioxide molecules contains one mole of carbon atoms weighing 12.011 grams and two moles of oxygen atoms weighing 2×15.999 grams for a total molecular mass of 44.009 grams per mole of carbon dioxide molecules.

The mass of a sample of a compound can be converted to moles by dividing its mass by the molar mass of the compound. This process is similar to the unit conversion discussed in the introduction to Standard Set 3. The number of particles in the sample is determined by multiplying the number of moles by Avogadro's number. The volume of an ideal or a nearly ideal gas at a fixed temperature and pressure is proportional to the number of moles. Students should be able to calculate the number of moles of a gas from its volume by using the relationship that at standard temperature and pressure (0°C and 1 atmosphere), one mole of gas occupies a volume of 22.4 liters.

> **3. e.** *Students know* how to calculate the masses of reactants and products in a chemical reaction from the mass of one of the reactants or products and the relevant atomic masses.

Atoms are neither created nor destroyed in a chemical reaction. When the chemical reaction is written as a balanced expression, it is possible to calculate the mass of any one of the products or of any one of the reactants if the mass of just one reactant or product is known.

Students can be taught how to use balanced chemical equations to predict the mass of any product or reactant. Teachers should emphasize that the coefficients in the balanced chemical equation are mole quantities, not masses. Here is an example: How many grams of water will be obtained by combining 5.0 grams of hydrogen gas with an excess of oxygen gas, according to the following balanced equation?

$$2H_2 + O_2 \rightarrow 2H_2O$$

This calculation is often set up algebraically, for example, as

$$5 \text{ g H}_2 \times \frac{1 \text{ mole H}_2}{2 \text{g H}_2} \times \frac{2 \text{ mole H}_2O}{2 \text{ mole H}_2} \times \frac{18 \text{ g H}_2O}{1 \text{ mole H}_2O} = ? \text{ g H}_2O$$

and can be easily completed by direct calculation and unit cancellation (dimensional analysis). Students should learn to recognize that the coefficients in the balanced equations refer to moles rather than to mass.

3. f.* *Students know* how to calculate percent yield in a chemical reaction.

Students can use a balanced equation for a chemical reaction to calculate from the given masses of reactants the masses of the resulting products. The masses so calculated represent the theoretical 100 percent conversion of reactants to products. When one or more of the products are weighed, the masses are often less than the theoretical 100 percent yield. One explanation is that the reaction may not go to completion and therefore will not convert all the reactants to products. A second possible explanation is that product material is lost in the separation and purification process. A third reason is that alternative reactions are taking place, leading to products different from those predicted. Percent yield is a standard way to compare actual and theoretical yields. It is defined as

$$\frac{\text{actual yield of product obtained in an experiment}}{\text{theoretical yield as determined by stoichiometric calculations}} \times 100\%$$

3. g.* *Students know* how to identify reactions that involve oxidation and reduction and how to balance oxidation-reduction reactions.

Oxidation of an element is defined as an increase in oxidation number, or a loss of electrons. *Reduction of an element* is defined as a decrease in oxidation number, or a gain in electrons. The assignment of oxidation numbers, or oxidation states, is a bookkeeping device. This process may be defined as the charge assigned to an atom, as though all the electrons in each bond were located on the more electronegative atom in the bond. Students should be taught how to assign oxidation numbers to atoms in free elements and in compounds. Lists of rules for this procedure are commonly found in chemistry textbooks.

In many important chemical reactions, elements change their oxidation states. These changes are called *redox,* or *oxidation-reduction reactions.* Respiration and photosynthesis are common examples with which students are familiar.

Any chemical reaction in which electrons are transferred from one substance to another is an oxidation-reduction reaction. Transfer can be determined by checking the oxidation states of atoms in reactants and products. A single displacement reaction, such as zinc metal in a copper sulfate solution, is a typical redox reaction because the zinc atoms lose two electrons and the copper ions gain two electrons. Redox reactions may be balanced by ensuring that the number of electrons lost in one part of the reaction equals the number of electrons gained in another part of the reaction. This principle and the conservation of atoms will hold true in any balanced oxidation/reduction reaction.

Students should be taught how to use the half-reaction method for balancing redox equations. This method divides the reaction into an oxidation portion and a reduction portion, both of which are balanced separately and then combined into an overall equation with no net change in number of electrons.

STANDARD SET 4. Gases and Their Properties

Standard Set 4 requires a knowledge of applicable physical concepts and sufficient skills in mathematical problem solving to describe (1) gases at the molecular level; (2) the behavior of gases; and (3) the measurable properties of gases. As background, students should know the physical states of matter and their properties and know how mixtures, especially homogeneous mixtures, differ from pure substances. Students should know that temperature measures how hot a system is, regardless of the system's size, and that heat flows from a region of higher temperature to one of lower temperature. Familiarity with the Celsius and Kelvin temperature scales is necessary. Knowledge of the motion of particles (and objects) and of kinetic energy is also required. Students should have the mathematical background to recognize and use directly and inversely proportional relationships. The ability to solve algebraic equations with several given quantities and one unknown is essential.

The main assumptions of the kinetic molecular theory should be covered in presenting this material, and a connection should be made to the behavior of ideal gases. Students are often more comfortable with studying the volume and temperature of a gas, but the concept of pressure also needs to be dealt with thoroughly. With knowledge of the kinetic molecular theory, students should see how motions and collisions of particles produce measurable properties, such as pressure.

> **4. The kinetic molecular theory describes the motion of atoms and molecules and explains the properties of gases. As a basis for understanding this concept:**
>
> **a.** *Students know* the random motion of molecules and their collisions with a surface create the observable pressure on that surface.

Fluids consist of molecules that freely move past each other in random directions. Intermolecular forces hold the atoms or molecules in liquids close to each other. Gases consist of tiny particles, either atoms or molecules, spaced far apart from each other and reasonably free to move at high speeds, near the speed of sound. In the study of chemistry, gases and liquids are considered fluids.

Pressure is defined as force per unit area. The force in fluids comes from collisions of atoms or molecules with the walls of the container. Air pressure is created by the weight of the gas in the atmosphere striking surfaces. Gravity pulls air molecules toward Earth, the surface that they strike. Water pressure can be understood in the same fashion, but the pressures are much greater because of the greater density of water. Pressure in water increases with depth, and pressure in air decreases with altitude. However, pressure is felt equally in all directions in fluids because of the random motion of the molecules.

> **4. b.** *Students know* the random motion of molecules explains the diffusion of gases.

Another result of the kinetic molecular theory is that gases diffuse into each other to form homogeneous mixtures. An excellent demonstration of diffusion is the white ammonium chloride ring formed by simultaneous diffusion of ammonia vapor and hydrogen chloride gas toward the middle of a glass tube. The white ring forms nearer the region where hydrogen chloride was introduced, illustrating both diffusion and the principle that heavier gases have a slower rate of diffusion.

> **4. c.** *Students know* how to apply the gas laws to relations between the pressure, temperature, and volume of any amount of an ideal gas or any mixture of ideal gases.

A fixed number of moles *n* of gas can have different values for pressure *P*, volume *V*, and temperature *T*. Relationships among these properties are defined for an ideal gas and can be used to predict the effects of changing one or more of these properties and solving for unknown quantities. Students should know and be able to use the three gas law relationships summarized in Table 1, "Gas Law Relationships."

Table 1
Gas Law Relationships

Expression of gas laws	Fixed values	Variable relationships	Form for calculations
PV = constant	n, T	Inverse	$P_1 V_1 = P_2 V_2$
V/T = constant	n, P	Direct	$V_1/T_1 = V_2/T_2$
P/T = constant	n, V	Direct	$P_1/T_1 = P_2/T_2$

The first expression of the gas law shown in Table 1 is sometimes taught as Boyle's law and the second as Charles's law, according to the historical order of their discovery. They are both simpler cases of the more general ideal gas law introduced in Standard 4.h in this section. For a fixed number of moles of gas, a combined gas law has the form PV/T = constant, or $P_1 V_1/T_1 = P_2 V_2/T_2$. This law is useful in calculations where *P*, *V*, and *T* are changing. By placing a balloon over the mouth of an Erlenmeyer flask, the teacher can demonstrate that volume divided by temperature equals a constant. When the flask is heated, the balloon inflates; when the flask is cooled, the balloon deflates.

> **4. d.** *Students know* the values and meanings of standard temperature and pressure (STP).

Standard temperature is 0°C, and standard pressure (STP) is 1 atmosphere (760 mm Hg). These standards are an agreed-on set of conditions for gases against which to consider other temperatures and pressures. When volumes of gases are being compared, the temperature and pressure must be specified. For a fixed mass of gas at a specified temperature and pressure, the volume is also fixed.

> **4. e.** *Students know* how to convert between the Celsius and Kelvin temperature scales.

Some chemical calculations require an absolute temperature scale, called the Kelvin scale (K), for which the coldest possible temperature is equal to zero. There are no negative temperatures on the Kelvin scale. In theory if a sample of any material is cooled as much as possible, the lowest temperature that can be reached is 0 K, experimentally determined as equivalent to −273.15°C. The Kelvin scale starts with absolute zero (0 K) because of this theoretical lowest temperature limit. A Kelvin temperature is always 273.15 degrees greater than an equivalent Celsius temperature, but a Kelvin temperature is specified without the degree symbol. The magnitude of one unit of change in the K scale is equal to the magnitude of one unit of change on the °C scale.

> **4. f.** *Students know* there is no temperature lower than 0 Kelvin.

The kinetic molecular theory is the basis for understanding heat and temperature. The greater the atomic and molecular motion, the greater the observed temperature of a substance. If all atomic and molecular motion stopped, the temperature of the material would reach an absolute minimum. This minimum is absolute zero, or −273.15°C. The third law of thermodynamics states that this temperature can never be reached. Experimental efforts to create very low temperatures have resulted in lowering the temperature of objects to within a fraction of a degree of absolute zero.

> **4. g.*** *Students know* the kinetic theory of gases relates the absolute temperature of a gas to the average kinetic energy of its molecules or atoms.

The value of the average kinetic energy for an ideal gas is directly proportional to its Kelvin temperature. Average kinetic energy can be related to changes in pressure and volume as a function of temperature. At 0 K all motion in an ideal monatomic gas ceases, meaning that the average kinetic energy equals zero.

4. h.* *Students know* how to solve problems by using the ideal gas law in the form $PV = nRT$.

The relationships among pressure, volume, and temperature for a fixed mass of gas can be expressed as the ideal gas law, $PV = nRT$, where n represents moles and R represents the universal gas constant, which is 0.0821 liter-atmosphere per mole-Kelvin (this unit can be abbreviated as (L atm)/(mol K) or as L atm mol^{-1} K^{-1}).

4. i.* *Students know* how to apply Dalton's law of partial pressures to describe the composition of gases and Graham's law to predict diffusion of gases.

It is important to distinguish clearly between *diffusion* and *effusion*. *Diffusion* is the process by which separate atoms or molecules intermingle as a result of random motion. *Effusion* is the process by which gas molecules pass from one container to another at lower pressure through a very small opening.

Graham's law states that the rates of effusion of two gases at the same temperature and pressure are inversely proportional to the square roots of their molar masses. Graham's law also approximately applies to rates of diffusion for gases, although diffusion is a more complicated process to describe than effusion.

Dalton's law of partial pressures states that total pressure in a gas-filled container is equal to the sum of the partial pressures of the component gases. This law can be introduced by showing that ideal gases have properties based solely on the number of moles present in the sample, without regard to the chemical identities of the gas particles involved.

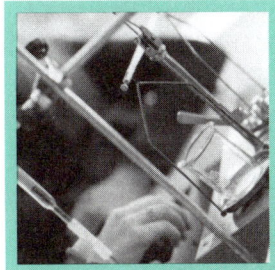

STANDARD SET 5. Acids and Bases

Students who learn the concepts in this standard set will be able to understand and explain aqueous acid–base reactions, properties of acids and bases, and pH as a measure of acidity and basicity. Careful thought should be given to how this standard set best fits with the other chemistry standards. It may be desirable to cover Standard Set 6, "Solutions," in this section first so that students will know about the aqueous dissolving process and about concentration calculations and units. Familiarity with Standard Set 9, "Chemical Equilibrium," in this section may help students to conceptualize the strengths of acids and bases.

Students should be able to represent and balance chemical reactions. They should also be able to interpret periodic trends in electronegativity for the upper two rows of the periodic table (see Standard Set 1, "Atomic and Molecular Structure," in this section). Hydrogen with its low electronegativity easily forms the positive hydrogen ion, H$^+$. Students need to know the charge and formula of the hydroxide ion, OH$^-$. With knowledge of polar covalent bonding, students should be able to distinguish between two important types of neutral molecular compounds: those that dissolve in an aqueous solution and remain almost completely as neutral molecules and those that dissolve in an aqueous solution and partially or almost completely produce charged ions (see Standard Sets 1 and 2 in this section).

Students should be able to compare the three descriptions of acids and bases—the Arrhenius, Brønsted-Lowry, and Lewis acid–base definitions—and recognize electron lone pairs on Lewis dot structures of molecules (see Standard Set 2, "Chemical Bonds," in this section). To calculate pH, students should understand and be able to use base-10 logarithms and antilogarithms and know how to obtain logarithms by using a calculator. Students should become proficient at converting between pH, pOH, [H^+], and [OH^-].

> **5. Acids, bases, and salts are three classes of compounds that form ions in water solutions. As a basis for understanding this concept:**
>
> **a.** *Students know* the observable properties of acids, bases, and salt solutions.

Comparing and contrasting the properties of acids and bases provide a context for understanding their behavior. Some observable properties of acids are that they taste sour; change the color of litmus paper from blue to red; indicate acidic values on universal indicator paper; react with certain metals to produce hydrogen gas; and react with metal hydroxides, or bases, to produce water and a salt. Some observable properties of bases are that basic substances taste bitter or feel slippery; change the color of litmus paper from red to blue; indicate basic values on universal indicator paper; and react with many compounds containing hydrogen ions, or acids, to produce water and a salt.

These properties can be effectively demonstrated by using extracted pigment from red cabbage as an indicator to analyze solutions of household ammonia and white vinegar at various concentrations. When the indicator is added, basic solutions turn green, and acidic solutions turn red. Students can also use universal indicator solutions to test common household substances. **Students need to follow established safety procedures while conducting experiments.**

> **5. b.** *Students know* acids are hydrogen-ion-donating and bases are hydrogen-ion-accepting substances.

According to the Brønsted-Lowry acid–base definition, acids donate hydrogen ions, and bases accept hydrogen ions. Acids that are formed from the nonmetals found in the first and second rows of the periodic table easily dissociate to produce hydrogen ions because these nonmetals have a large electronegativity compared with that of hydrogen. Once students know that acids and bases have different effects on the same indicator, they are ready to deepen their understanding of acid–base behavior at the molecular level. Examples and studies of chemical reactions should be used to demonstrate these definitions of acids and bases.

5. c. *Students know* strong acids and bases fully dissociate and weak acids and bases partially dissociate.

Acids dissociate by donating hydrogen ions, and bases ionize by dissociating to form hydroxide ions (from a hydroxide salt) or by accepting hydrogen ions. Some acids and bases either dissociate or ionize almost completely, and others do so only partially. Nearly complete dissociation is strong; partial dissociation is weak. The strength of an acid or a base can vary, depending on such conditions as temperature and concentration.

5. d. *Students know* how to use the pH scale to characterize acid and base solutions.

The pH scale measures the concentrations of hydrogen ions in solution and the acidic or basic nature of the solution. The scale is not linear but logarithmic, meaning that at pH 2, for example, the concentration of hydrogen ions is ten times greater than it is at pH 3. The pH scale ranges from below 0 (very acidic) to above 14 (very basic). Students should learn that pH values less than 7 are considered acidic and those greater than 7 are considered basic.

5. e.* *Students know* the Arrhenius, Brønsted-Lowry, and Lewis acid–base definitions.

Other acid–base definitions beside Brønsted-Lowry (see Standard 5.b in this section) are the Arrhenius and Lewis definitions. An Arrhenius base must contain hydroxide, such as in the chemical KOH. NH_3 does not contain hydroxide and therefore would not be a base according to the Arrhenius definition. However, because NH_3 accepts hydrogen ions, it would be a base according to the Brønsted-Lowry definition. A Lewis acid is an electron pair receptor, and a Lewis base is an electron pair donor. Using the Lewis definition extends the concept of acid–base reactions to nonaqueous systems. The compound BF_3, for example, is an acid because it accepts a lone electron pair, but BF_3 would not be an acid according to the Brønsted-Lowry definition because the hydrogen ion is not present.

5. f.* *Students know* how to calculate pH from the hydrogen-ion concentration.

The pH scale is defined as $-\log_{10}[H^+]$, where $[H^+]$ is the hydrogen-ion concentration in moles per liter of solution. Students should be taught to convert between hydrogen-ion concentration and pH.

5. g.* *Students know* buffers stabilize pH in acid–base reactions.

A buffer is a solution that stabilizes H^+ concentration levels. Such a solution may release hydrogen ions as pH rises or consume hydrogen ions as pH decreases. A critically important, but extremely complex example is the equilibria between carbon dioxide, carbonic acid, bicarbonate, carbonate, and solid calcium carbonate that keeps the world's oceans at a nearly constant pH of about 8. Another measure students should learn is pOH, which is the negative logarithm of the OH⁻ concentration expressed in moles per liter of solution. The sum of pH and pOH is always 14.0 for a given solution at 25°C.

STANDARD SET 6. Solutions

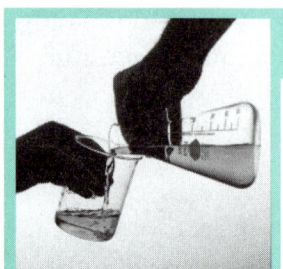

As background for this standard set, students must know the physical states of matter and the corresponding properties. Specifically, the properties of the liquid state are important because most discussions will focus on liquid solutions. Students also need an understanding of molecules and ions and sufficient mathematical skills (especially an understanding of unit conversions) to describe solutions at the ionic or molecular level, describe the dissolution process, and quantify solution concentrations. Moreover, students should be able to calculate mass and volume in a variety of units and work with ratios, percentages, and moles (see Standard Set 3, "Conservation of Matter and Stoichiometry").

The sequencing of instruction regarding these standards deserves special attention and consideration. The knowledge of mixtures (especially of homogeneous mixtures), which will be acquired from the study of these standards, can be helpful in reinforcing the concepts of electronegativity, covalent bonding, shapes of molecules, and ionic compounds (see Standard Set 1, "Atomic and Molecular Structure," and Standard Set 2, "Chemical Bonds"). Some of these standards may also be helpful prerequisite knowledge for understanding Standard Set 5, "Acids and Bases." These standards also help prepare students to study the separation or purification of solution components as covered in Standard Set 9, "Chemical Equilibrium." Toward that end, instruction in these standards should include the concept of solubility as an equilibrium between solid and solute forms of a substance and the study of reactions that lead to precipitate formation.

> **6. Solutions are homogeneous mixtures of two or more substances. As a basis for understanding this concept:**
>
> **a.** *Students know* the definitions of *solute* and *solvent*.

Simple solutions are homogeneous mixtures of two substances. A *solute* is the dissolved substance in a solution, and a *solvent* is, by quantity, the major component in the solution.

> **6. b.** *Students know* how to describe the dissolving process at the molecular level by using the concept of random molecular motion.

The kinetic molecular theory as applied to gases can be extended to explain how the solute and solvent particles are in constant random motion. The kinetic energy of this motion causes diffusion of the solute into the solvent, resulting in a homogeneous solution. When a solid is in contact with a liquid, at least some small degree of dissolution always occurs. The equilibrium concentration of solute in solvent will depend on the surface interactions between the molecules of solute and solvent. Equilibrium is reached when all competing processes are in balance. Those processes include the tendency for dissolved molecules to spread randomly in the solvent and the competing strength of the bonds and other forces among solute molecules, among solvent molecules, and between solute and solvent molecules. When salts dissolve in water, positive and negative ions are separated and surrounded by polar water molecules.

> **6. c.** *Students know* temperature, pressure, and surface area affect the dissolving process.

In a liquid solvent, solubility of gases and solids is a function of temperature. Students should have experience with reactions in which precipitates are formed or gases are released from solution, and they should be taught that the concentration of a substance that appears as solid or gas must exceed the solubility of the solvent.

Increasing the temperature usually increases the solubility of solid solutes but always decreases the solubility of gaseous solutes. An example of a solid ionic solute compound that decreases in solubility as the temperature increases is Na_2SO_4. An example of one that increases in solubility as the temperature increases is $NaNO_3$. The solubility of a gas in a liquid is directly proportional to the pressure of that gas above the solution. It is important to distinguish solubility equilibrium from rates of dissolution. Concepts of equilibrium describe only how much solute will dissolve at equilibrium, not how quickly this process will occur.

> **6. d.** *Students know* how to calculate the concentration of a solute in terms of grams per liter, molarity, parts per million, and percent composition.

All concentration units listed previously are a measure of the amount of solute compared with the amount of solution. Grams per liter represent the mass of solute divided by the volume of solution. *Molarity* describes moles of solute divided by liters of solution. Students can calculate the number of moles of dissolved solute and divide by the volume in liters of the total solution, yielding units of moles per liter. Parts per million, which is a ratio of one part of solute to one million parts of solvent, is usually applied to very dilute solutions. Percent composition is the ratio of one part of solvent to one hundred parts of solvent and is expressed as a percent. To calculate parts per million and percent composition, students determine the mass of solvent and solute and then divide the mass of the solute by the total mass of the solution. This number is then multiplied by 10^6 and expressed as parts per million (ppm) or by 100 and expressed as a percent.

> **6. e.*** *Students know* the relationship between the molality of a solute in a solution and the solution's depressed freezing point or elevated boiling point.

The physical properties of the freezing point and boiling point of a solution are directly proportional to the concentration of the solution in molality. *Molality* is similar to *molarity* except that molality expresses moles of solute dissolved in a kilogram of solvent rather than in a liter of solution. In other words, *molality* is the amount of solute present divided by the amount of solvent present. Molality is a convenient measure because it does not depend on volume and therefore does not change with temperature. Sometimes physical properties change with concentration of solute; for example, salt, such as sodium chloride or calcium chloride, is sprinkled on icy roads to lower the freezing point of water and melt the snow or ice. The freezing point is lowered, or depressed, in proportion to the amount of salt dissolved.

> **6. f.*** *Students know* how molecules in a solution are separated or purified by the methods of chromatography and distillation.

Chromatography is a powerful, commonly used method to separate substances for analysis, including DNA, protein, and metal ions. The principle takes advantage of a moving solvent and a stationary substrate to induce separation. A useful, interesting example is paper chromatography. In this laboratory technique a mixture of solutes, such as ink dyes, is applied to a sheet of chromatographic paper. One end of the paper is dipped in a solvent that moves (wicks) up or along the paper. Solutes (the various ink dyes, for example) separate into bands of colors. Solutes with great affinity for the paper move little, those with less move more, and those with very little affinity may travel with the leading edge of the solvent.

Mixtures can sometimes be separated by distillation, which capitalizes on differences in the forces holding molecules in a liquid state. Crude oil, for example, is processed by commercial refineries (by a catalytic reaction called "cracking") and separated by heat distillation to give a variety of commercial products, from highly volatile kerosene and gasoline to heavier oils used to lubricate engines or to heat homes.

STANDARD SET 7. Chemical Thermodynamics

Students should know the relationship between heat and temperature and understand the concepts of kinetic energy and motion at the molecular level. They should also know how to differentiate between chemical and physical change and understand that both types of change entail the loss of free energy, a process that results in increased stability. The standards marked with an asterisk in this section require students to know the concepts presented in the physics section of this chapter, Standard Set 3, "Heat and Thermodynamics." Those standards provide a foundation for understanding the kinetic molecular model and introduce students to concepts of heat and entropy.

> **7. Energy is exchanged or transformed in all chemical reactions and physical changes of matter. As a basis for understanding this concept:**
>
> **a.** *Students know* how to describe temperature and heat flow in terms of the motion of molecules (or atoms).

Temperature is a measure of the average kinetic energy of molecular motion in a sample. *Heat* is energy transferred from a sample at higher temperature to one at lower temperature. Often, heat is described as flowing from the system to the surroundings or from the surroundings to the system. The system is defined by its boundaries, and the surroundings are outside the boundaries, with "the universe" frequently considered as the surroundings.

> **7. b.** *Students know* chemical processes can either release (exothermic) or absorb (endothermic) thermal energy.

Endothermic processes absorb heat, and their equations can be written with heat as a reactant. *Exothermic processes* release heat, and their equations can be written with heat as a product. The net heat released to or absorbed from the surroundings comes from the making and breaking of chemical bonds during a reaction. Students understand and relate heat to the internal motion of the atoms and molecules. They also understand that breaking a bond always requires energy and that making a bond almost always releases energy. The amount of energy per bond depends on the strength of the bond.

The potential energy of the reaction system may be plotted for the different reaction stages: reactants, transition states, and products. This plot will show reactants at lower potential energy than products for an endothermic reaction and reactants at higher potential energy than products for an exothermic reaction. A higher energy transition state usually exists between the reactant and product energy states that affect the reaction rate covered in Standard Set 8, "Reaction Rates," in this section.

> **7. c.** *Students know* energy is released when a material condenses or freezes and is absorbed when a material evaporates or melts.

Physical changes are accompanied by changes in internal energy. Changes of physical state either absorb or release heat. Evaporation and melting require energy to overcome the bonds of attractions in the corresponding liquid or solid state. Condensation and freezing release heat to the surroundings as internal energy is reduced and bonds of attraction are formed.

> **7. d.** *Students know* how to solve problems involving heat flow and temperature changes, using known values of specific heat and latent heat of phase change.

Qualitative knowledge that students gained by mastering the previous standards will help them to solve problems related to the heating or cooling of a substance over a given temperature range. *Specific heat* is the energy needed to change the temperature of one gram of substance by one degree Celsius. The unit of specific heat is joule/gram-degree.

During phase changes, energy is added or removed without a corresponding temperature change. This phenomenon is called *latent* (or *hidden*) *heat*. There is a latent heat of fusion and a latent heat of vaporization. The unit of latent heat is joule/gram or kilojoule/mole. Students should be able to diagram the temperature changes that occur when ice at a temperature below zero is heated to superheated steam, which has temperatures above 100°C.

> **7. e.*** *Students know* how to apply Hess's law to calculate enthalpy change in a reaction.

As samples of elements combine to make compounds, heat may be absorbed from the environment. If one mole of a compound is formed from elements and all substances begin and end at 25°C and are under standard atmospheric pressure, the heat absorbed during the compound's synthesis is known as its *standard enthalpy of formation*, H_f. If heat is not absorbed but released, H_f is negative. Values for the enthalpy of formation of thousands of compounds are available in reference books.

Hess's law states that if a chemical reaction is carried out in any imaginable series of steps, the net enthalpy change (heat absorbed) in the reaction is the sum of the enthalpy changes for the individual steps. For example, a reaction can be imagined to proceed in just two steps: first, making its reactants into elements, and second, making those elements into products. The enthalpy change in a reaction $aA + bB = cC + dD$ is the quantity ΔH_r° defined as "the heat absorbed when a moles of chemical A and b moles of B react to make c moles of C and d moles of D." Through the use of Hess's law and the two steps of the example, the enthalpy change for this reaction can be expressed in terms of the enthalpy of formation H_f° of each of the chemicals: the enthalpy change is the (weighted) sum of the heat of formation for each of the products minus the sum for each of the reactants, or

$$\Delta H_r^\circ = [c\,H_f^\circ(C) + d\,H_f^\circ(D)] - [a\,H_f^\circ(A) + b\,H_f^\circ(B)]$$

By using the balanced equation for a chemical reaction and the enthalpy of formation for each chemical, students can calculate the heat absorbed or released when a given quantity of a reactant is consumed.

7. f.* *Students know* how to use the Gibbs free energy equation to determine whether a reaction would be spontaneous.

Endothermic and exothermic reactions can be spontaneous under standard conditions of temperature and pressure. Therefore, releasing heat and going to a lower energy state cannot be the only force driving chemical reactions. The tendency to disorder, or entropy, is the other driving force. A convenient conceptual way to account for the balance between these two driving forces, enthalpy changes and changes in disorder, was developed by J. Willard Gibbs and is called the Gibbs free-energy change ΔG.

In the Gibbs free-energy equation (shown below), ΔH is the change in enthalpy, T is the Kelvin temperature, and ΔS is the change in entropy. (Note that students may need to be introduced to the basics of entropy, as presented in Standard Set 3 for physics in this chapter).

$$\Delta G = \Delta H - T\Delta S$$

The Gibbs free-energy change is used to predict in which direction a reaction will proceed. A negative value for the Gibbs free-energy change predicts the formation of products (a spontaneous reaction); a positive value predicts, or favors, reactants (a nonspontaneous reaction). Standard values for the Gibbs free energy of elements and compounds at a specified temperature are available in tables.

STANDARD SET 8. Reaction Rates

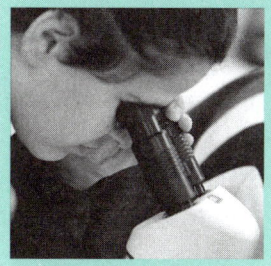

To describe rates of chemical reactions, factors affecting rates, and the energy changes involved, students need to know that chemical reactions consume reactants and form products (see Standard Set 3, "Conservation of Matter and Stoichiometry," in this section). Students should be able to explain chemical reactions at the molecular level and know how kinetic energy at the molecular level is measured by temperature (see Standard Sets 3 and 7 in this section). Students have acquired knowledge of pressure and volume relationships for gases (see Standard Set 4, "Gases and Their Properties," in this section) and can plot potential energy versus course of reaction for endothermic and exothermic reactions (see Standard Set 7, "Chemical Thermodynamics," in this section). The ability to calculate rates of change from slopes of lines and curves is required.

8. Chemical reaction rates depend on factors that influence the frequency of collision of reactant molecules. As a basis for understanding this concept:

   a. *Students know* the rate of reaction is the decrease in concentration of reactants or the increase in concentration of products with time.

Students may have an intuitive idea that reaction rate is a measure of how fast reactions proceed, but a quantitative measure for reaction rate also is needed. For example, explosive reactions are very fast, as are many biological reactions in the cell; and other reactions, such as iron rusting, are very slow. *Reaction rate* is defined as the rate of decrease in concentration of reactants or as the rate of increase in concentration of products, and these reciprocal changes form a balanced equation that reflects the conservation of matter. Students can see from the balanced equation that as the reaction proceeds, the concentration of reactants must decrease, and the concentration of products must increase in proportion to their mole ratios.

> **8. b.** *Students know* how reaction rates depend on such factors as concentration, temperature, and pressure.

Concentration, temperature, and pressure should be emphasized because they are major factors affecting the collision of reactant molecules and, thus, affecting reaction rates. Increasing the concentration of reactants increases the number of collisions per unit time. Increasing temperature (which increases the average kinetic energy of molecules) also increases the number of collisions per unit time. Though the collision rate modestly increases, the greater kinetic energy dramatically increases the chances of each collision leading to a reaction (e.g., the Arrhenius effect). Increasing pressure increases the reaction rate only when one or more of the reactants or products are gases. With gaseous reactants, increasing pressure is the same as increasing concentration and results in an elevated reaction rate.

> **8. c.** *Students know* the role a catalyst plays in increasing the reaction rate.

A *catalyst* increases the rate of a chemical reaction without taking part in the net reaction. A catalyst lowers the energy barrier between reactants and products by promoting a more favorable pathway for the reaction. Surfaces often play important roles as catalysts for many reactions. One reactant might be temporarily held on the surface of a catalyst. There the bonds of the reactant may be weakened, allowing another substance to react with it more quickly. Living systems speed up life-dependent reactions with biological catalysts called *enzymes*. Catalysts are used in automobile exhaust systems to reduce the emission of smog-producing unburned hydrocarbons.

> **8. d.*** *Students know* the definition and role of activation energy in a chemical reaction.

Even in a spontaneous reaction, reactants are usually required to pass through a transition state that has a higher energy than either the reactants or the products. The additional energy, called the *activation energy,* or the *activation barrier,* is related to such factors as strength of bonding within the reactants. The more energy required to go from reactants to activated transition complex, the higher the activation barrier, and the slower a reaction will be. Catalysts speed up rates by lowering the activation barrier along the reaction pathway between products and reactants.

STANDARD SET 9. Chemical Equilibrium

Chemical equilibrium is a dynamic state. To understand the factors affecting equilibrium and to write expressions used to quantify a state of equilibrium, students will need a thorough knowledge of reaction rates (see Standard Set 8, "Reaction Rates," in this section) and of chemical thermodynamics (see Standard Set 7, "Chemical Thermodynamics," in this section). Changes in heat accompanying chemical reactions and spontaneity of chemical reactions are key topics. Students should be able to identify the physical states of substances undergoing chemical reactions and use knowledge of substances at the atomic and molecular levels (see Standard Set 3, "Conservation of Matter and Stoichiometry," in this section).

Students need to know how gases respond to changes in pressure and volume (see Standard Set 4, "Gases and Their Properties," in this section). Students need to be able to calculate concentration and molarity for solutions, particularly for aqueous solutions (see Standard Set 6, "Solutions," in this section). Students familiar with acid–base reactions (see Standard Set 5, "Acids and Bases," in this section) and precipitation reactions (see Standard Set 6 in this section) will have an advantage in learning the concept of equilibrium.

To calculate equilibrium constants, students should be able to balance chemical equations (see Standard Set 3 in this section), work readily with concentration and pressure units (see Standard Sets 4 and 6 in this section), and use exponents in mathematical calculations.

Students often have difficulty in understanding that equilibrium, which is a dynamic process, occurs when no net changes in a product or reactant concentration take place. An analogy can be made with a pair of escalators operating between two floors. If the same number of people go up as go down in a ten-minute interval, the rate of people moving up equals the rate of people moving down. Overall, any extra people arriving on one floor are canceled out by others leaving the floor. Therefore, the number of people on each floor will be constant over time, and the population of the two floors is in dynamic equilibrium. This analogy can be extended to a chemical reaction by considering that if the number of moles produced in one direction of the reaction is the same as the number consumed in the opposite direction, then the reaction has reached a state of dynamic equilibrium. Students will learn that when a stress is applied to a chemical reaction in equilibrium, a shift will occur to partly relieve the stress.

> **9. Chemical equilibrium is a dynamic process at the molecular level. As a basis for understanding this concept:**
>   **a.** *Students know* how to use Le Chatelier's principle to predict the effect of changes in concentration, temperature, and pressure.

Le Chatelier's principle can be introduced by emphasizing the balanced nature of an equilibrium system. If an equilibrium system is stressed or disturbed, the system will respond (change or shift) to partially relieve or undo the stress. A new equilibrium will eventually be established with a new set of conditions. When the stress is applied, the reaction is no longer at equilibrium and will shift to regain equilibrium. For instance, if the concentration of a reactant in a system in dynamic equilibrium is decreased, products will be consumed to produce more of that reactant. Students need to remember that heat is a reactant in endothermic reactions and a product in exothermic reactions. Therefore, increasing temperature will shift an endothermic reaction, for example, to the right to regain equilibrium. Students should note that any endothermic chemical reaction is exothermic in the reverse direction.

Pressure is proportional to concentration for gases; therefore, for chemical reactions that have a gaseous product or reactant, pressure affects the system as a whole. Increased pressure shifts the equilibrium toward the smaller number of moles of gas, alleviating the pressure stress. If both sides of the equilibrium have an equal number of moles of gas, increasing pressure does not affect the equilibrium. Adding an inert gas, such as argon, to a reaction will not change the partial pressures of the reactant or product gases and therefore will have no effect on the equilibrium.

> **9. b.** *Students know* equilibrium is established when forward and reverse reaction rates are equal.

Forward and reverse reactions at equilibrium are going on at the same time and at the same rate, causing overall concentrations of each reactant and product to remain constant over time.

> **9. c.*** *Students know* how to write and calculate an equilibrium constant expression for a reaction.

Because the concentrations of substances in a system at chemical equilibrium are constant over time, chemical expressions related to each concentration will also be constant. Here is a general equation for a reaction at equilibrium:

$a\text{A} + b\text{B} \leftrightarrow c\text{C} + d\text{D}$

The general expression for the equilibrium constant of a chemical reaction is K_{eq}, defined at a particular temperature, often 25°C. Its formula is

$$K_{eq} = \frac{[\text{C}]^c [\text{D}]^d}{[\text{A}]^a [\text{B}]^b}$$

When K_{eq} is being calculated, only gaseous substances and aqueous solutions are considered. Equilibrium concentrations of products, in moles per liter, are in the numerator, and equilibrium concentrations of reactants are in the denominator. The exponents are the corresponding coefficients from the balanced chemical equation. A large K_{eq} means the forward reaction goes almost to completion; that is, little reverse reaction occurs. A very small K_{eq} means the reverse reaction goes almost to completion, or little forward reaction occurs. The solubility product constant K_{sp} is the equilibrium constant for salts in solution.

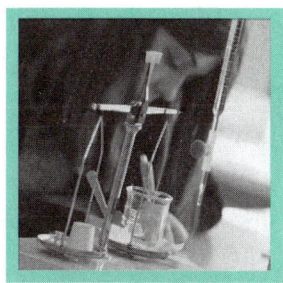

STANDARD SET 10. Organic and Biochemistry

A solid understanding of chemical and biological concepts is required to describe the versatility with which carbon atoms form molecules and to illustrate the structure of organic and biological molecules and the polymers they form. An understanding of these concepts is also necessary to be able to name organic molecules and to identify organic functional groups. Students also need to know the material in atomic and molecular structure (see Standard Set 1, "Atomic and Molecular Structure," in this section), especially as the concepts pertain to carbon and nearby nonmetals and to electronegativity.

Chemical bonds (see Standard Set 2, "Chemical Bonds," in this section), especially the topics of covalent bonding and Lewis dot structures, should also be well understood. Students should know the importance of biochemical compounds—proteins, carbohydrates, and nucleic acids—and understand their roles in living organisms. They must be familiar with single, double, and triple bonds to name organic molecules correctly (see Standard Set 2 in this section) and with the covalent bonding and electronegativity of nitrogen and oxygen to identify functional groups of organic molecules and to understand amino acids and proteins (see Standard Sets 1 and 2 in this section).

> **10. The bonding characteristics of carbon allow the formation of many different organic molecules of varied sizes, shapes, and chemical properties and provide the biochemical basis of life. As a basis for understanding this concept:**
>
> **a.** *Students know* large molecules (polymers), such as proteins, nucleic acids, and starch, are formed by repetitive combinations of simple subunits.

Students can readily visualize large molecules called polymers as consisting of repetitive and systematic combinations of smaller, simpler groups of atoms, including carbon. All polymeric molecules, including biological molecules, such as proteins, nucleic acids, and starch, are made up of various unique combinations of a relatively small number of chemically simple subunits. For example, starch is a polymer made from a large number of simple sugar molecules joined together.

> **10. b.** *Students know* the bonding characteristics of carbon that result in the formation of a large variety of structures ranging from simple hydrocarbons to complex polymers and biological molecules.

Building on what they learned in grade eight about the unique bonding characteristics of carbon, students explore in greater depth the incredible diversity of carbon-based molecules. They are reminded that, given carbon's four bonding electrons and four vacancies available to form bonds, carbon is able to form stable covalent bonds—single or multiple—with other carbon atoms and with atoms of other elements.

Students learn how the presence of single, double, and triple bonds determines the geometry of carbon-based molecules. The variety of these molecules is enormous: over 16 million carbon-containing compounds are known. The compounds range from simple hydrocarbon molecules (e.g., methane and ethane) to complex organic polymers and biological molecules (e.g., proteins) and include many manufactured polymers used in daily life (e.g., polyester, nylon, and polyethylene).

> **10. c.** *Students know* amino acids are the building blocks of proteins.

Proteins are large single-stranded polymers often made up of thousands of relatively small subunits called *amino acids.* The bond attaching two amino acids, known as the *peptide bond,* is identical for any pair of amino acids. The chemical composition of the amino acid itself varies. Variation in composition and ordering of amino acids gives protein molecules their unique properties and shapes. These properties and shapes define the protein's functions, many of which are essential to the life of an organism. The blueprint for building the protein molecules is deoxyribonucleic acid (DNA). Biotechnology is advancing rapidly as more is learned about DNA, amino acid sequences, and the shapes and functions of proteins.

> **10. d.*** *Students know* the system for naming the ten simplest linear hydrocarbons and isomers that contain single bonds, simple hydrocarbons with double and triple bonds, and simple molecules that contain a benzene ring.

Organic molecules can be simple or extremely complex. The naming system for these molecules, however, is relatively straightforward and reflects the composition and structure of each molecule. Each name is made up of a prefix and a suffix. The prefix tells the number of carbon atoms in the longest continuous sequence of the molecule, and the suffix indicates the kind of bond between carbon atoms. For example, the four simplest hydrocarbon molecules are methane, ethane, propane, and butane. The prefixes, *meth-, eth-, prop-,* and *but-* refer to one, two, three, and four carbons, respectively. The *-ane* ending indicates that there are only carbon-carbon single bonds. *Ene* endings are used for double bonds and *-yne* for triple bonds. Benzene, C_6H_6, is a flat hexagonally shaped molecule of six carbon atoms bonded to each other. Many compounds can be built by substitutions on straight-chain hydrocarbons and benzene rings.

> **10. e.*** *Students know* how to identify the functional groups that form the basis of alcohols, ketones, ethers, amines, esters, aldehydes, and organic acids.

Organic molecules are grouped into classes based on patterns of bonding between carbon and noncarbon atoms (e.g., nitrogen and oxygen). Groups based on unique patterns of bonding are called *functional groups.* Examples of these groups are alcohols, ketones, ethers, amines, esters, aldehydes, and organic (carboxylic) acids.

> **10. f.*** *Students know* the R-group structure of amino acids and know how they combine to form the polypeptide backbone structure of proteins.

Amino acid molecules have a well-known structure, and all contain a side chain called an *R-group*. Differences in the R-group are the basis for differences between the amino acids. Bonding two amino acids creates a *dipeptide*, bonding three creates a *tripeptide*, and adding more creates a polymer called a *polypeptide*. Polypeptides made biologically are called *proteins*.

STANDARD SET 11. Nuclear Processes

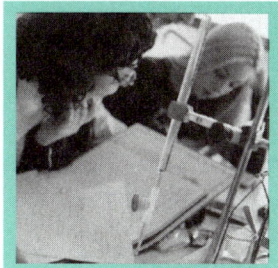

This section requires a knowledge of chemical and physical concepts and sufficient mathematical skills to describe the nucleus and its subatomic particles. Topics covered are nuclear reactions and their accompanying changes in energy and forms of radiation and quantification of radioactive decay as a function of time. Students should know about nuclear structure and properties, the mass and charge of the proton and neutron, and the use of the periodic table to determine the number of protons in an atom's nucleus (see Standard Set 1, "Atomic and Molecular Structure," in this section). Students should also know how to calculate and use percentages to determine the amount of radioactive substance remaining after a time interval of disintegration.

Students should be introduced to this standard set with a review of the nucleus and its constituent protons and neutrons. Simple hydrogen, deuterium, and tritium nuclei can be used to introduce and define *isotopes*. Students should be reminded of the difference between an element's average atomic mass and the mass number for a specific isotope. They should already know that electrons and protons attract each other as do any particles of opposite charge, but they probably do not know what holds the protons and neutrons in the nucleus together. Teachers should introduce students to the strong nuclear force and explain how it holds protons and neutrons together and how it can overcome the repulsion between charged protons at very close distances. Students should also be introduced to quarks as the constituents of protons and neutrons.

> **11. Nuclear processes are those in which an atomic nucleus changes, including radioactive decay of naturally occurring and human-made isotopes, nuclear fission, and nuclear fusion. As a basis for understanding this concept:**
>
> **a.** *Students know* protons and neutrons in the nucleus are held together by nuclear forces that overcome the electromagnetic repulsion between the protons.

The nucleus is held together by the strong nuclear force. The strong nuclear force acts between protons, between neutrons, and between protons and neutrons but has a limited range comparable to the size of an atomic nucleus. The nuclear force is able to overcome the mutual electrostatic repulsion of the protons only when the protons and neutrons are near each other as they are in the nucleus of an atom.

> **11. b.** *Students know* the energy release per gram of material is much larger in nuclear fusion or fission reactions than in chemical reactions. The change in mass (calculated by $E = mc^2$) is small but significant in nuclear reactions.

Two major types of nuclear reactions are fusion and fission. In *fusion* reactions two nuclei come together and merge to form a heavier nucleus. In *fission* a heavy nucleus splits apart to form two (or more) lighter nuclei. The binding energy of a nucleus depends on the number of neutrons and protons it contains. A general term for a proton or a neutron is a *nucleon.* In both fusion and fission reactions, the total number of nucleons does not change, but large amounts of energy are released as nucleons combine into different arrangements. This energy is one million times more than energies involved in chemical reactions.

> **11. c.** *Students know* some naturally occurring isotopes of elements are radioactive, as are isotopes formed in nuclear reactions.

Sometimes atoms with the same number of protons in the nucleus have different numbers of neutrons. These atoms are called *isotopes* of an element. Both naturally occurring and human-made isotopes of elements can be either stable or unstable. Less stable isotopes of one element, called *parent isotopes,* will undergo radioactive decay, transforming to more stable isotopes of another element, called *daughter products,* which can also be either stable or radioactive. For a radioactive isotope to be found in nature, it must either have a long half-life, such as potassium-40, uranium-238, uranium-235, or thorium-232, or be the daughter product, such as radon-222, of a parent with a long half-life, such as uranium-238.

> **11. d.** *Students know* the three most common forms of radioactive decay (alpha, beta, and gamma) and know how the nucleus changes in each type of decay.

Radioactive isotopes transform to more stable isotopes, emitting particles from the nucleus. These particles are helium-4 nuclei (alpha radiation), electrons or positrons (beta radiation), or high-energy electromagnetic rays (gamma radiation). Isotopes of elements that undergo alpha decay produce other isotopes with two less protons and two less neutrons than the original isotope. Uranium-238, for instance, emits an alpha particle and becomes thorium-234.

Isotopes of elements that undergo beta decay produce elements with the same number of nucleons but with one more proton or one less proton. For example, thorium-234 beta decays to protactinium-234, which then beta decays to uranium-234. Alpha and beta decay are ionizing radiations with the potential to damage surrounding materials. After alpha and beta decay, the resulting nuclei often emit high-energy photons called *gamma rays*. This process does not change the number of nucleons in the nucleus of the isotope but brings about a lower energy state in the nucleus.

> **11. e.** *Students know* alpha, beta, and gamma radiation produce different amounts and kinds of damage in matter and have different penetrations.

Alpha, beta, and gamma rays are *ionizing radiations,* meaning that those rays produce tracks of ions of atoms and molecules when they interact with materials. For all three types of rays, the energies of particles emitted in radioactive decay are typically for each particle on the order of 1MeV, equal to 1.6×10^{-13} joule, which is enough energy to ionize as many as half a million atoms.

Alpha particles have the shortest ranges, and matter that is only a few millimeters thick will stop them. They will not penetrate a thick sheet of paper but will deposit all their energy along a relatively short path, resulting in a high degree of ionization along that path.

Beta particles have longer ranges, typically penetrating matter up to several centimeters thick. Those particles are electrons or positrons (the antimatter electron), have one unit of either negative or positive electric charge, and are approximately 1/2000 of the mass of a proton. These high-energy electrons have longer ranges than alpha particles and deposit their energy along longer paths, spreading the ionization over a greater distance in the material.

Gamma rays can penetrate matter up to several meters thick. Gamma rays are high-energy photons that have no electric charge and no rest mass (the structural energy of the particle). They will travel unimpeded through materials until they strike an electron or the nucleus of an atom. The gamma ray's energy will then be either completely or partially absorbed, and neighboring atoms will be ionized. Therefore, these three types of radiation interact with matter by losing energy and ionizing surrounding atoms.

Alpha radiation is dangerous if ingested or inhaled. For example, radon-222, a noble gas element, is a naturally occurring hazard in some regions. Living organisms or sensitive materials can be protected from ionizing radiation by shielding them and increasing their distance from radiation sources.

Because many people deeply fear and misunderstand radioactivity, chemistry teachers should address and explore the ability of each form of radiation to penetrate matter and cause damage. Students may be familiar with radon detection devices, similar to smoke detectors, found in many homes. Discussion of biological and health effects of ionizing radiation can inform students about the risks and benefits of nuclear reactions. Videos can be used in the classroom to show demonstrations of the penetrating ability of alpha, beta, and gamma radiation through paper, aluminum, and lead or through other dense substances of varying thicknesses. Geiger counter measurements can be used to record radiation data. The order of penetrating ability, from greatest to least, is gamma > beta > alpha, and this order is the basis for assessing proper shielding of radiation sources for safety.

There are a number of naturally occurring sources of ionizing radiation. One is potassium-40, which can be detected easily in potash fertilizer by using a Geiger counter. The other is background cosmic and alpha radiation from radon. This radiation can be seen in cloud chambers improvised in the classroom.

> **11. f.*** *Students know* how to calculate the amount of a radioactive substance remaining after an integral number of half-lives have passed.

Radioactive decay transforms the initial (parent) nuclei into more stable (daughter) nuclei with a characteristic half-life. The *half-life* is the time it takes for one-half of a given number of parent atoms to decay to daughter atoms. One-half of the remaining parent atoms will then decay to produce more daughter atoms in the next half-life period. It is possible to predict only the proportion, not the individual number, of parent atoms that will undergo decay. Therefore, after one half-life, 50 percent of the initial parent nuclei remain; after two half-lives, 25 percent; and so forth. The intensity of radiation from a radioactive source is related to the half-life and to the original number of radioactive atoms present.

> **11. g.*** *Students know* protons and neutrons have substructures and consist of particles called *quarks*.

Just as atoms consist of subatomic nuclear particles (protons and neutrons), so do protons and neutrons have constituent particles called *quarks*. Quarks come in six different types, but only two types are involved in ordinary matter: the *up quark* and the *down quark*.

For enrichment beyond the content in the standard, the following description is included: Protons and neutrons each contain three quarks. A proton consists of two up quarks and one down quark. A neutron consists of two down quarks and one up quark. Quarks have fractional electric charges; therefore, the charge of an up quark is $+\frac{2}{3}$ units, and the charge of a down quark is $-\frac{1}{3}$ units. It is believed that it is not possible to isolate a free quark. All the common matter in the material world is made up of just three fundamental particles: the up quark, the down quark, and the electron.

Biology/Life Sciences

Living organisms appear in many variations, yet there are basic similarities among their forms and functions. For example, all organisms require an outside source of energy to sustain life processes; all organisms demonstrate patterns of growth and, in many cases, senescence, the process of becoming old; and the continuity of all species requires reproduction. All organisms are constructed from the same types of macromolecules (proteins, nucleic acids, lipids) and inherit a deoxyribonucleic acid (DNA) genome from a parent or parents. DNA is always transcribed to yield ribonucleic acid (RNA), which is translated through the use of a nearly universal genetic code. Environmental factors frequently regulate and influence the expression of specific genes.

Biologists study life at many levels, and the biology standards for grades nine through twelve reflect these studies. *Organisms* are part of an ecosystem and have complex relationships with other organisms and the physical environment. Ecologists study these populations and communities, and many are deeply interested in the physical and behavioral adaptations of organisms. Evolutionary biologists share these interests because the fitness of an organism is a manifestation of these adaptations. *Adaptations* are traits subject to the rules of inheritance; therefore, genetics and evolutionary biology are closely allied fields.

Physiologists study whole body systems or organs. For example, a neurophysiologist focuses primarily on the nervous system. Cell biologists study the details of how cells and organelles work, considering such weighty matters as how cytoskeletal elements segregate chromosomes during mitosis, how proteins are sorted to different compartments of the cell, and how receptors in the cell membrane communicate with factors that regulate gene expression. Many cell biologists also consider themselves to be developmental biologists, molecular geneticists, or biochemists. There are many connections between all the fields and different ways of viewing life.

Biology textbooks typically start with a review of chemistry and energetics; therefore, California students will be able to make good use of their study of the content standards for "Chemistry of Living Systems" in the eighth grade. The principles of cellular biology, including respiration and photosynthesis, are usually taught next, followed by instruction in molecular and Mendelian genetics. Population genetics and evolution follow naturally from the study of genetics and lead to a discussion of diversity of form and physiology. The teaching culminates with ecology, a subject that draws on each of the preceding topics. The teaching comes full circle because ecology is also a starting point for students in lower elementary school grade levels.

STANDARD SET 1. Cell Biology

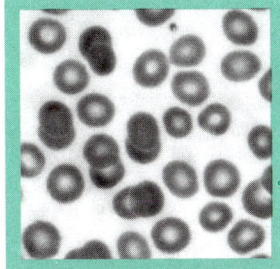

The first knowledge of cells came from the work of an English scientist, Robert Hooke, who in 1665 used a primitive microscope to study thin sections of cork and called the boxlike cavities he saw "cells." Antony van Leeuwenhoek later observed one-celled "animalcules" in pond water, but not until the 1830s did Theodor Schwann view cartilage tissue in which he discovered cells resembling plant cells. He published the theory that cells are the basic unit of life. Rudolf Virchow used the work of Schwann and Matthias Schleiden to advance the cell theory, presenting the concept that plants and animals are made of cells that contain fluid and nuclei and arise from preexisting cells.

After the cell theory was established, detailed study of cell structure and function depended on the improvement of microscopes and on techniques for preparing specimens for observation. It is now understood that cells in plants and animals contain genes to control chemical reactions needed for survival and organelles to perform those reactions. Living organisms may consist of one cell, as in bacteria, or of many cells acting in a coordinated and cooperative manner, as in plants, animals, and fungi. All cells have at least three structures in common: genetic material, a cell or plasma membrane, and cytoplasm.

> 1. **The fundamental life processes of plants and animals depend on a variety of chemical reactions that occur in specialized areas of the organism's cells. As a basis for understanding this concept:**
>
>   a. *Students know* cells are enclosed within semipermeable membranes that regulate their interaction with their surroundings.

The plasma membrane consists of two layers of lipid molecules organized with the polar (globular) heads of the molecules forming the outside of the membrane and the nonpolar (straight) tails forming the interior of the membrane. Protein molecules embedded within the membrane move about relative to one another in a fluid fashion. Because of its dynamic nature the membrane is sometimes referred to as *the fluid mosaic model* of membrane structure.

Cell membranes have three major ways of taking in or of regulating the passage of materials into and out of the cell: simple diffusion, carrier-facilitated diffusion, and active transport. Osmosis of water is a form of diffusion. Simple diffusion and carrier-facilitated diffusion do not require the expenditure of chemical bond energy, and the net movement of materials reflects a concentration gradient or a voltage gradient or both. Active transport requires free energy, in the form of either chemical bond energy or a coupled concentration gradient, and permits the net transport or "pumping" of materials against a concentration gradient.

> **1. b.** *Students know* enzymes are proteins that catalyze biochemical reactions without altering the reaction equilibrium and the activities of enzymes depend on the temperature, ionic conditions, and the pH of the surroundings.

Almost all *enzymes* are protein catalysts made by living organisms. Enzymes speed up favorable (spontaneous) reactions by reducing the activation energy required for the reaction, but they are not consumed in the reactions they promote. To demonstrate the action of enzymes on a substrate, the teacher can use liver homogenate or yeast as a source of the enzyme catalase and hydrogen peroxide as the substrate. The effect of various environmental factors, such as pH, temperature, and substrate concentration, on the rate of reaction can be investigated. These investigations should encourage student observation, recording of qualitative and quantitative data, and graphing and interpretation of data.

> **1. c.** *Students know* how prokaryotic cells, eukaryotic cells (including those from plants and animals), and viruses differ in complexity and general structure.

All living cells are divided into one of two groups according to their cellular structure. *Prokaryotes* have no membrane-bound organelles and are represented by the Kingdom Monera, which in modern nomenclature is subdivided into the Eubacteria and Archaea. *Eukaryotes* have a complex internal structure that allows thousands of chemical reactions to proceed simultaneously in various organelles. *Viruses* are not cells; they consist of only a protein coat surrounding a strand of genetic material, either RNA or DNA.

> **1. d.** *Students know* the central dogma of molecular biology outlines the flow of information from transcription of ribonucleic acid (RNA) in the nucleus to translation of proteins on ribosomes in the cytoplasm.

DNA, which is found in the nucleus of eukaryotes, contains the genetic information for encoding proteins. The DNA sequence specifying a specific protein is copied (transcribed) into messenger RNA (mRNA), which then carries this message out of the nucleus to the ribosomes located in the cytoplasm. The mRNA message is then translated, or converted, into the protein originally coded for by the DNA.

> **1. e.** *Students know* the role of the endoplasmic reticulum and Golgi apparatus in the secretion of proteins.

There are two types—rough and smooth—of endoplasmic reticulum (ER), both of which are systems of folded sacs and interconnected channels. *Rough ER* synthesizes proteins, and *smooth ER* modifies or detoxifies lipids. Rough ER produces new proteins, including membrane proteins. The proteins to be exported from the cell are moved to the Golgi apparatus for modification, packaged in vesicles, and transported to the plasma membrane for secretion.

> **1. f.** *Students know* usable energy is captured from sunlight by chloroplasts and is stored through the synthesis of sugar from carbon dioxide.

Photosynthesis is a complex process in which visible sunlight is converted into chemical energy in carbohydrate molecules. This process occurs within chloroplasts and specifically within the thylakoid membrane (light-dependent reaction) and the stroma (light-independent reaction). During the light-dependent reaction, water is oxidized and light energy is converted into chemical bond energy generating ATP, NADPH + H$^+$, and oxygen gas.[†] During the light-independent reaction (Calvin cycle), carbon dioxide, ATP, and NADPH + H$^+$ react, forming phosphoglyceraldehyde, which is then converted into sugars. By using a microscope with appropriate magnification, students can see the chloroplasts in plant cells (e.g., lettuce, onion) and photosynthetic protists (e.g., euglena).

Students can prepare slides of these cells themselves, an activity that provides a good opportunity to see the necessity for well-made thin sections of specimens and for correct staining procedures. Commercially prepared slides are also available. By observing prepared cross sections of a leaf under a microscope, students can see how a leaf is organized structurally and think about the access of cells to light and carbon dioxide during photosynthesis. The production of oxygen from photosynthesis can be demonstrated and measured quantitatively with a volumeter, which can collect oxygen gas from the illuminated leaves of an aquatic plant, such as elodea. By varying the distance between the light source and the plant, teachers can demonstrate intensities of the effects of various illumination. To eliminate heat as a factor, the teacher can place a heat sink, such as a flat-sided bottle of water, between the plant and light source to absorb or dissipate unwanted heat.

> **1. g.** *Students know* the role of the mitochondria in making stored chemical-bond energy available to cells by completing the breakdown of glucose to carbon dioxide.

Mitochondria consist of a matrix where three-carbon fragments originating from carbohydrates are broken down (to CO_2 and water) and of the cristae where ATP is produced. Cell respiration occurs in a series of reactions in which fats, proteins, and carbohydrates, mostly glucose, are broken down to produce carbon dioxide, water, and energy. Most of the energy from cell respiration is converted into ATP, a substance that powers most cell activities.

> **1. h.** *Students know* most macromolecules (polysaccharides, nucleic acids, proteins, lipids) in cells and organisms are synthesized from a small collection of simple precursors.

Many of the large carbon compound molecules necessary for life (e.g., polysaccharides, nucleic acids, proteins, and lipids) are polymers of smaller monomers. Polysaccharides are composed of monosaccharides; proteins are composed of amino

---

[†]ATP is adenosine triphosphate, and NADPH is reduced nicotinamide adenine dinucleotide phosphate.

acids; lipids are composed of fatty acids, glycerol, and other components; and nucleic acids are composed of nucleotides.

> **1. i.*** *Students know* how chemiosmotic gradients in the mitochondria and chloroplast store energy for ATP production.

Enzymes called *ATP synthase,* located within the thylakoid membranes in chloroplasts and cristae membranes in mitochondria, synthesize most ATP within cells. The thylakoid and cristae membranes are impermeable to protons except at pores that are coupled with the ATP synthase. The potential energy of the proton concentration gradient drives ATP synthesis as the protons move through the ATP synthase pores. The proton gradient is established by energy furnished by a flow of electrons passing through the electron transport system located within these membranes.

> **1. j.*** *Students know* how eukaryotic cells are given shape and internal organization by a cytoskeleton or cell wall or both.

The *cytoskeleton,* which gives shape to and organizes eukaryotic cells, is composed of fine protein threads called *microfilaments* and thin protein tubes called *microtubules.* Cilia and *flagella* are composed of microtubules arranged in the 9 + 2 arrangement, in which nine pairs of microtubules surround two single microtubules. The rapid assembly and disassembly of microtubules and microfilaments and their capacity to slide past one another enable cells to move, as observed in white blood cells and amoebae, and also account for movement of organelles within the cell. Students can observe prepared slides of plant mitosis in an onion root tip to see the microtubules that make up the spindle apparatus. Prepared slides of white fish blastula reveal animal spindle apparatus and centrioles, both of which are composed of microtubules.

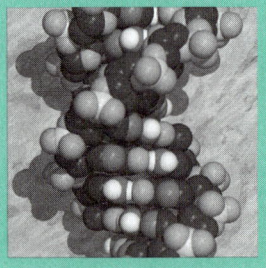

STANDARD SET 2. Genetics (Meiosis and Fertilization)

Students should know that organisms reproduce offspring of their own kind and that organisms of the same species resemble each other. Students have been introduced to the idea that some characteristics can be passed from parents to offspring and that individual variations appear among offspring and in the broader population. Understanding genetic variation requires mastery of the fundamentals of sex cell formation and the steps to reorganize and redistribute genetic material during defined stages in the cell cycle.

Students should understand the difference between asexual cell reproduction (*mitosis*) and the formation of male or female gamete cells (*meiosis*). Sexual reproduction initially requires the production of haploid eggs and haploid sperm, a process occurring in humans within the female ovary and the male testis. These haploid cells unite in fertilization and produce the *diploid zygote,* or fertilized cell.

The mechanisms involved in synapsis and movement of chromosomes during meiosis bring about the halving of the chromosome numbers for the production of the haploid male or female gamete cells from the original diploid parent cell and different combinations of parental genes. The exchange of chromosomal segments between homologous chromosomes (crossing over) revises the association of genes on the chromosomes and contributes to increased diversity. Any change in genetic constitution through mutation, crossing over, or chromosome assortment during meiosis promotes genetic variation in a population.

> **2. Mutation and sexual reproduction lead to genetic variation in a population. As a basis for understanding this concept:**
>   **a.** *Students know* meiosis is an early step in sexual reproduction in which the pairs of chromosomes separate and segregate randomly during cell division to produce gametes containing one chromosome of each type.

Haploid gamete production through meiosis involves two cell divisions. During meiosis prophase I, the homologous chromosomes are paired, a process that abets the exchange of chromosome parts through breakage and reunion. The second meiotic division parallels the mechanics of mitosis except that this division is not preceded by a round of DNA replication; therefore, the cells end up with the haploid number of chromosomes. (The nucleus in a haploid cell contains one set of chromosomes.) Four haploid nuclei are produced from the two divisions that characterize meiosis, and each of the four resulting cells has different chromosomal constituents (components). In the male all four become sperm cells. In the female only one becomes an egg, while the other three remain small degenerate polar bodies and cannot be fertilized.

Chromosome models can be constructed and used to illustrate the segregation taking place during the phases of mitosis (covered initially in Standard 1.e for grade seven in Chapter 4) and meiosis. Commercially available optical microscope slides also show cells captured in mitosis (onion root tip) or meiosis (Ascaris blastocyst cells), and computer and video animations are also available.

> **2. b.** *Students know* only certain cells in a multicellular organism undergo meiosis.

Only special diploid cells, called *spermatogonia* in the testis of the male and *oogonia* in the female ovary, undergo meiotic divisions to produce the haploid sperm and haploid eggs.

> **2. c.** *Students know* how random chromosome segregation explains the probability that a particular allele will be in a gamete.

The steps in meiosis involve random chromosome segregation, a process that accounts for the probability that a particular allele will be packaged in any given gamete. This process allows for genetic predictions based on laws of probability that pertain to genetic sortings. Students can create a genetic chart and mark alternate traits on chromosomes, one expression coming from the mother and another expression coming from the father. Students can be shown that partitions of the chromosomes are controlled by chance (are random) and that separation (segregation) of chromosomes during *karyokinesis* (division of the nucleus) leads to the random sequestering of different combinations of chromosomes.

> **2. d.** *Students know* new combinations of alleles may be generated in a zygote through the fusion of male and female gametes (fertilization).

Once gametes are formed, the second half of sexual reproduction can take place. In this process a diploid organism is reconstituted from two haploid parts. When a sperm is coupled with an egg, a fertilized egg (zygote) is produced that contains the combined genotypes of the parents to produce a new allelic composition for the progeny. Genetic charts can be used to illustrate how new combinations of alleles may be present in a zygote through the events of meiosis and the chance union of gametes. Students should be able to read the genetic diploid karyotype, or chromosomal makeup, of a fertilized egg and compare the allelic composition of progeny with the genotypes and phenotypes of the parents.

> **2. e.** *Students know* why approximately half of an individual's DNA sequence comes from each parent.

Chromosomes are composed of a single, very long molecule of double-stranded DNA and proteins. Genes are defined as segments of DNA that code for polypeptides (proteins). During fertilization half the DNA of the progeny comes from the gamete of one parent, and the other half comes from the gamete of the other parent.

> **2. f.** *Students know* the role of chromosomes in determining an individual's sex.

The normal human somatic cell contains 46 chromosomes, of which 44 are pairs of homologous chromosomes and 2 are sex chromosomes. Females usually carry two X chromosomes, and males possess one X and a smaller Y chromosome. Combinations of these two sex chromosomes determine the sex of the progeny.

> **2. g.** *Students know* how to predict possible combinations of alleles in a zygote from the genetic makeup of the parents.

When the genetic makeups of potential parents are known, the possible assortments of alleles in their gametes can be determined for each genetic locus. Two parental gametes will fuse during fertilization, and with all pair-wise combinations of their gametes considered, the possible genetic makeups of progeny can then be predicted.

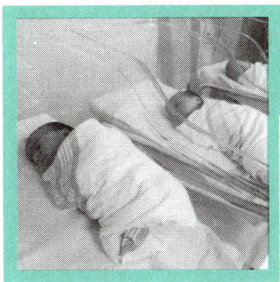

STANDARD SET 3. Genetics (Mendel's Laws)

Breeding of plants and animals has been an active technology for thousands of years, but the science of heredity is linked to the genetics pioneer Gregor Mendel. He studied phenotypic traits of various plants, especially those of peas. (A *phenotypic trait* is the physical appearance of a trait in an organism). From the appearance of these traits in different generations of growth, he was able to infer their *genotypes* (the genetic makeup of an organism with respect to a trait) and to speculate about the genetic makeup and method of transfer of the hereditary units from one generation to the next. (Probability analysis is now used to predict probable progeny phenotypes from various parental genetic crosses.) The genetic basis for Mendel's laws of segregation and independent assortment is apparent from genetic outcomes of crosses.

> **3. A multicellular organism develops from a single zygote, and its phenotype depends on its genotype, which is established at fertilization. As a basis for understanding this concept:**
>
> **a.** *Students know* how to predict the probable outcome of phenotypes in a genetic cross from the genotypes of the parents and mode of inheritance (autosomal or X-linked, dominant or recessive).

Monohybrid crosses, including autosomal dominant alleles, autosomal recessive alleles, incomplete dominant alleles, and X-linked alleles, can be used to indicate the parental genotypes and phenotypes. The possible gametes derived from each parent are based on genotypic ratios and can be used to predict possible progeny. The predictive (probabilistic) methods for determining the outcome of genotypes and phenotypes in a genetic cross can be introduced by using Punnett Squares and probability mathematics.

Teachers should review the process of writing genotypes and help students translate genotypes into phenotypes. Teachers should emphasize dominant, recessive, and incomplete dominance as the students advance to an explanation of monohybrid crosses illustrating human conditions characterized by autosomal recessive alleles, such as albinism, cystic fibrosis, Tay-Sachs, and phenylketonuria (PKU). These disorders can be contrasted with those produced by possession of just one autosomal dominant allele, conditions such as Huntington disease, dwarfism, and neurofibromatosis. This basic introduction can be followed with examples of incomplete dominance, such as seen in the comparisons of straight, curly, and wavy hair or in the expression of intermediate flower colors in snapdragon plants.

Sex-linked characteristics that are found only on the X chromosome should also be considered, and students should reflect on how this mode of transmission can cause the exclusive or near-exclusive appearance in males of color blindness, hemophilia, fragile-X syndrome, and sex-linked muscular dystrophy.

3. b. *Students know* the genetic basis for Mendel's laws of segregation and independent assortment.

Mendel deduced that for each characteristic, an organism inherits two genes, one from each parent. When the two alleles differ, the dominant allele is expressed, and the recessive allele remains hidden. Two genes or alleles separate (segregate) during gamete production in meiosis, resulting in the sorting of alleles into separate gametes (the law of segregation). Students can be shown how to diagram Mendel's explanation for how a trait present in the parental generation can appear to vanish in the first filial (F1) generation of a monohybrid cross and then reappear in the following second filial (F2) generation.

Students should be told that alternate versions of a gene at a single locus are called *alleles*. Students should understand Mendel's deduction that for each character, an organism inherits two genes, one from each parent. From this point students should realize that if the two alleles differ, the dominant allele, if there is one, is expressed, and the recessive allele remains hidden. Students should recall that the two genes, or alleles, separate (segregate) during gamete production in meiosis and that this sorting of alleles into separate gametes is the basis for the law of segregation. This law applies most accurately when genes reside on separate chromosomes that segregate out at random, and it often does not apply or is a poor predictor for combinations and frequencies of genes that reside on the same chromosome. Students can study various resources that describe Mendel's logic and build models to illustrate the laws of segregation and independent assortment.

3. c.* *Students know* how to predict the probable mode of inheritance from a pedigree diagram showing phenotypes.

Students should be taught how to use a pedigree diagram showing phenotypes to predict the mode of inheritance.

3. d.* *Students know* how to use data on frequency of recombination at meiosis to estimate genetic distances between loci and to interpret genetic maps of chromosomes.

Students should be able to interpret genetic maps of chromosomes and manipulate genetic data by using standard techniques to relate recombination at meiosis to estimate genetic distances between loci.

STANDARD SET 4. Genetics (Molecular Biology)

All cells contain DNA as their genetic material. The role of DNA in organisms is twofold: first, to store and transfer genetic information from one generation to the next, and second, to express that genetic information in the synthesis of proteins. By controlling protein synthesis, DNA controls the structure and function of all cells. The complexity of an organism determines whether it may have several hundred to more than twenty thousand proteins as a part of its makeup.

Proteins are composed of a sequence of amino acids linked by peptide bonds (see Standard 10.c for chemistry in this chapter). The identity, number, and sequence of the amino acids in a protein give each protein its unique structure and function. Twenty types of amino acids are commonly employed in proteins, and each can appear many times in a single protein molecule. The proper sequence of amino acids in a protein is translated from an RNA sequence that is itself encoded in the DNA.

> **4. Genes are a set of instructions encoded in the DNA sequence of each organism that specify the sequence of amino acids in proteins characteristic of that organism. As a basis for understanding this concept:**
>
> a. *Students know* the general pathway by which ribosomes synthesize proteins, using tRNAs to translate genetic information in mRNA.

DNA does not leave the cell nucleus, but messenger RNA (mRNA), complementary to DNA, carries encoded information from DNA to the ribosomes (transcription) in the cytoplasm. (The *ribosomes* translate mRNAs to make protein.) Freely floating amino acids within the cytoplasm are bonded to specific transfer RNAs (tRNAs) that then transport the amino acid to the mRNA now located on the ribosome. As a ribosome moves along the mRNA strand, each mRNA codon, or sequence of three nucleotides specifying the insertion of a particular amino acid, is paired in sequence with the anticodon of the tRNA that recognizes the sequence. Each amino acid is added, in turn, to the growing polypeptide at the specified position.

After learning about transcription and translation through careful study of expository texts, students can simulate the processes on paper or with representative models. Computer software and commercial videos are available that illustrate animated sequences of transcription and translation.

> **4. b.** *Students know* how to apply the genetic coding rules to predict the sequence of amino acids from a sequence of codons in RNA.

The sequence of amino acids in protein is provided by the genetic information found in DNA. In prokaryotes, mRNA transcripts of a coding sequence are copied from the DNA as a single contiguous sequence. In eukaryotes, the initial RNA transcript, while in the nucleus, is composed of *exons,* sequences of nucleotides that carry useful information for protein synthesis, and *introns,* sequences that do not. Before leaving the nucleus, the initial transcript is processed to remove introns and splice exons together. The processed transcript, then properly called mRNA and carrying the appropriate codon sequence for a protein, is transported from the nucleus to the ribosome for translation.

Each mRNA has sequences, called *codons*, that are decoded three nucleotides at a time. Each codon specifies the addition of a single amino acid to a growing polypeptide chain. A *start codon* signals the beginning of the sequence of codons to be translated, and a *stop codon* ends the sequence to be translated into protein. Students can write out mRNA sequences with start and stop codons from a given DNA sequence and use a table of the genetic code to predict the primary sequences of proteins.

> **4. c.** *Students know* how mutations in the DNA sequence of a gene may or may not affect the expression of the gene or the sequence of amino acids in the encoded protein.

Mutations are permanent changes in the sequence of nitrogen-containing bases in DNA (see Standard 5.a in this section for details on DNA structure and nitrogen bases). Mutations occur when base pairs are incorrectly matched (e.g., *A* bonded to *C* rather than *A* bonded to *T*) and can, but usually do not, improve the product coded by the gene. Inserting or deleting base pairs in an existing gene can cause a mutation by changing the codon reading frame used by a ribosome. Mutations that occur in somatic, or nongerm, cells are often not detected because they cannot be passed on to offspring. They may, however, give rise to cancer or other undesirable cellular changes. Mutations in the germline can produce functionally different proteins that cause such genetic diseases as Tay-Sachs, sickle cell anemia, and Duchenne muscular dystrophy.

> **4. d.** *Students know* specialization of cells in multicellular organisms is usually due to different patterns of gene expression rather than to differences of the genes themselves.

Gene expression is a process in which a gene codes for a product, usually a protein, through transcription and translation. Nearly all cells in an organism contain the same DNA, but each cell transcribes only that portion of DNA containing the genetic information for proteins required at that specific time by that specific cell. The remainder of the DNA is not expressed. Specific types of cells may produce specific proteins unique to that type of cell.

> **4. e.** *Students know* proteins can differ from one another in the number and sequence of amino acids.

Protein molecules vary from about 50 to 3,000 amino acids in length. The types, sequences, and numbers of amino acids used determine the type of protein produced.

> **4. f.*** *Students know* why proteins having different amino acid sequences typically have different shapes and chemical properties.

The 20 different protein-making amino acids have the same basic structure: an amino group; an acidic (carboxyl) group; and an R, or radical group (see Standard

Set 10, "Organic and Biochemistry," in the chemistry section of this chapter). The protein is formed by the amino group of one amino acid linking to the carboxyl group of another amino acid. This bond, called the *peptide bond,* is repeated to form long molecular chains with the R groups attached along the polymer backbone.

The properties of these amino acids vary from one another because of both the order and the chemical properties of these R groups. Typically, the long protein molecule folds on itself, creating a three-dimensional structure related to its function. Structure, for example, may allow a protein to be a highly specific catalyst, or enzyme, able to position and hold other molecules. The R group of an amino acid consists of atoms that may include carbon, hydrogen, nitrogen, oxygen, and sulfur, depending on the amino acid. Amino acids containing sulfur sometimes play an important role of cross-linking and stabilizing polymer chains. Because of their various R groups, different amino acids vary in their chemical and physical properties, such as solubility in water, electrical charge, and size. These differences are reflected in the unique structure and function of each type of protein.

STANDARD SET 5. Genetics (Biotechnology)

Long before scientists identified DNA as the genetic material of cells, much was known about inheritance and the relationships between various characteristics likely to appear from one generation to the next. However, to comprehend clearly how the genetic composition of cells changes, students need to understand the structure and activity of nucleic acids.

Genetic recombination occurs naturally in sexual reproduction, viral infection, and bacterial transformation. These natural events change the genetic makeup of organisms and provide the raw materials for evolution. Natural selection determines the usefulness of the recombinants. In recombinant DNA technology specific pieces of DNA are recombined in the laboratory to achieve a specific goal. The scientist, rather than natural selection, then determines the usefulness of the recombinant DNA created.

> **5.** **The genetic composition of cells can be altered by incorporation of exogenous DNA into the cells. As a basis for understanding this concept:**
>
>   **a.** *Students know* the general structures and functions of DNA, RNA, and protein.

Nucleic acids are polymers composed of monomers called *nucleotides.* Each nucleotide consists of three subunits: a five-carbon pentose sugar, a phosphoric acid group, and one of four nitrogen bases. (For DNA these nitrogen bases are adenine, guanine, cytosine, or thymine.) DNA and RNA differ in a number of major ways. A DNA nucleotide contains a deoxyribose sugar, but RNA contains ribose sugar.

The nitrogen bases in RNA are the same as those in DNA except that thymine is replaced by uracil. RNA consists of only one strand of nucleotides instead of two as in DNA.

The DNA molecule consists of two strands twisted around each other into a double helix resembling a ladder twisted around its long axis. The outside, or uprights, of the ladder are formed by the two sugar-phosphate backbones. The rungs of the ladder are composed of pairs of nitrogen bases, one extending from each upright. In DNA these nitrogen bases always pair so that *T* pairs with *A,* and *G* pairs with *C.* This pairing is the reason DNA acts as a template for its own replication. RNA exists in many structural forms, many of which play different roles in protein synthesis. The mRNA form serves as a template during protein synthesis, and its codons are recognized by aminoacylated tRNAs. Protein and rRNA make up the structure of the ribosome.

Proteins are polymers composed of amino acid monomers (see Standard Set 10 for chemistry in this chapter). Different types of proteins function as enzymes and transport molecules, hormones, structural components of cells, and antibodies that fight infection. Most cells in an individual organism carry the same set of DNA instructions but do not use the entire DNA set all the time. Only a small amount of the DNA appropriate to the function of that cell is expressed. Genes are, therefore, turned on or turned off as needed by the cell, and the products coded by these genes are produced only when required.

> **5. b.** *Students know* how to apply base-pairing rules to explain precise copying of DNA during semiconservative replication and transcription of information from DNA into mRNA.

Enzymes initiate DNA replication by unzipping, or unwinding, the double helix to separate the two parental strands. Each strand acts as a template to form a complementary daughter strand of DNA. The new daughter strands are formed when complementary new nucleotides are added to the bases of the nucleotides on the parental strands. The nucleotide sequence of the parental strand dictates the order of the nucleotides in the daughter strands. One parental strand is conserved and joins a newly synthesized complementary strand to form the new double helix; this process is called *semiconservative replication.*

DNA replication is usually initiated by the separation of DNA strands in a small region to make a "replication bubble" in which DNA synthesis is primed. The DNA strands progressively unwind and are replicated as the replication bubble expands, and the two forks of replication move in opposite directions along the chromosome. At each of the diverging replication forks, the strand that is conserved remains a single, continuous "leading" strand, and the other "lagging" complementary strand is made as a series of short fragments that are subsequently repaired and ligated together.

Students may visualize DNA by constructing models, and they can simulate semiconservative replication by tracing the synthesis of the leading and lagging strands. The critical principles to teach with this activity are that two double-stranded DNA strands are the product of synthesis, that the process is semiconservative, that the antiparallel orientation of the strands requires repeated reinitiation on the lagging strand, and that the only information used during synthesis is specified by the base-pairing rules.

RNA is produced from DNA when a section of DNA (containing the nucleotide sequence required for the production of a specific protein) is transcribed. Only the template side of the DNA is copied. RNA then leaves the nucleus and travels to the cytoplasm, where protein synthesis takes place.

> **5. c.** *Students know* how genetic engineering (biotechnology) is used to produce novel biomedical and agricultural products.

Recombinant DNA contains DNA from two or more different sources. Bacterial plasmids and viruses are the two most common vectors, or carriers, by which recombinant DNA is introduced into a host cell. Restriction enzymes provide the means by which researchers cut DNA at desired locations to provide DNA fragments with "sticky ends." Genes, once identified, can be amplified either by cloning or by polymerase chain reactions, both of which produce large numbers of copies. The recombinant cells are then grown in large fermentation vessels, and their products are extracted from the cells (or from the medium if the products are secreted) and purified. Genes for human insulin, human growth hormone, blood clotting factors, and many other products have been identified and introduced into bacteria or other microorganisms that are then cultured for commercial production. Some agricultural applications of this technology are the identification and insertion of genes to increase the productivity of food crops and animals and to promote resistance to certain pests and herbicides, robustness in the face of harsh environmental conditions, and resistance to various viruses.

Students can model the recombinant DNA process by using paper models to represent eukaryotic complementary DNA (cDNA), the activity of different restriction enzymes, and ligation into plasmid DNA containing an antibiotic resistance gene and origin of DNA replication. To manipulate the modeled DNA sequences, students can use scissors (representing the activity of restriction enzymes) and tape (representing DNA ligase). If both strands are modeled on a paper tape, students can visualize how, in many cases, restriction enzymes make staggered cuts that generate "sticky ends" and how the ends must be matched during ligation.

> **5. d.*** *Students know* how basic DNA technology (restriction digestion by endonucleases, gel electrophoresis, ligation, and transformation) is used to construct recombinant DNA molecules.

In recombinant DNA technology DNA is isolated and exchanged between organisms to fulfill a specific human purpose. The desired gene is usually identified and extracted by using restriction enzymes, or *endonucleases*, to cut the DNA into fragments. Restriction enzymes typically cut *palindromic* portions of DNA, which read the same forward and backward, in ways that form sticky complementary ends. DNA from different sources, but with complementary sticky ends, can be joined by the enzyme DNA ligase, thus forming recombinant DNA.

DNA fragments of varying lengths can be separated from one another by *gel electrophoresis.* In this process the particles, propelled by an electric current, are moved through an agarose gel. Depending on the size, shape, and electrical charge of the particles, they will move at different rates through the gel and thus form bands of particles of similar size and charge. With appropriate staining, the various DNA fragments can then be visualized and removed for further analysis or recombination.

> **5. e.*** *Students know* how exogenous DNA can be inserted into bacterial cells to alter their genetic makeup and support expression of new protein products.

Bacteria can be induced to take up recombinant plasmids, a process called DNA transformation, and the plasmid is replicated as the bacteria reproduce. Recombinant bacteria can be grown to obtain billions of copies of the recombinant DNA. Commercially available kits containing all the necessary reagents, restriction enzymes, and bacteria are available for experiments in plasmid DNA transformation. Although the reagents and equipment can be expensive, various California corporations and universities have programs to make the cost more affordable, sometimes providing reagents and lending equipment.

Students should know that DNA transformation is a natural process and that horizontal DNA transfer is common in the wild. An example of how humans have manipulated genetic makeup is through the selective breeding of pets and of agricultural crops.

STANDARD SET 6. Ecology

Ecology is the study of relationships among living organisms and their interactions with the physical environment. These relationships are in a constant state of flux, and even small changes can cause effects throughout the ecosystem. Students in grades nine through twelve can be taught to think of ecology as changing relationships among the components of an ecosystem. Students also need to recognize that humans are participants in these ecosystem relationships, not just observers. A goal of classroom teaching should be to develop a strong scientific understanding of ecology to establish the basis for making informed and valid decisions.

> **6. Stability in an ecosystem is a balance between competing effects. As a basis for understanding this concept:**
>
> **a.** *Students know* biodiversity is the sum total of different kinds of organisms and is affected by alterations of habitats.

Biodiversity refers to the collective variety of living organisms in an ecosystem. This structure is influenced by alterations in habitat, including but not limited to climatic changes, fire, flood, and invasion by organisms from another system. The more biodiversity in an ecosystem, the greater its stability and resiliency. The best way for students to learn about ecology is to master the principles of the subject through careful study and then to make firsthand observations of ecosystems in action over time.

Although field trips are the ideal way to implement this process and should be encouraged, even career scientists often use models to study ecology. Local ecologists from government, private industry, or university programs may also be willing to serve as guest speakers in the classroom. Viewing the Internet's many virtual windows that show actual ecological experiments can also help students understand the scientific basis of ecology.

> **6. b.** *Students know* how to analyze changes in an ecosystem resulting from changes in climate, human activity, introduction of nonnative species, or changes in population size.

Analysis of change can help people to describe and understand what is happening in a natural system and, to some extent, to control or influence that system. Understanding different kinds of change can help to improve predictions of what will happen next. Changes in ecosystems often manifest themselves in predictable patterns of climate, seasonal reproductive cycles, population cycles, and migrations. However, unexpected disturbances caused by human intervention or the introduction of a new species, for example, may destabilize the often complex and delicate balance in an ecosystem.

Analyzing changes in an ecosystem can require complex methods and techniques because variation is not necessarily simple and may be interrelated with changes or trends in other factors. Rates and patterns of change, including trends, cycles, and irregularities, are essential features of the living world and are useful indicators of change that can provide data for analysis. Often it is important to analyze change over time, a process called *longitudinal analysis.*

> **6. c.** *Students know* how fluctuations in population size in an ecosystem are determined by the relative rates of birth, immigration, emigration, and death.

Fluctuations in the size of a population are often difficult to measure directly but may be estimated by measuring the relative rates of birth, death, immigration, and emigration in a population. The number of deaths and emigrations over time will decrease a population's size, and the number of births and immigrations over time will increase it. Comparing rates for death and emigration with those for birth and immigration will determine whether the population shows a net growth or a decline over time.

> **6. d.** *Students know* how water, carbon, and nitrogen cycle between abiotic resources and organic matter in the ecosystem and how oxygen cycles through photosynthesis and respiration.

Living things depend on nonliving things for life. At the organism level living things depend on natural resources, and at the molecular level, they depend on chemical cycles. Water, carbon, nitrogen, phosphorus, and other elements are recycled back and forth between organisms and their environments. Water, carbon, and nitrogen are necessary for life to exist. These chemicals are incorporated into plants (producers) by photosynthesis and nitrogen fixation and used by animals (consumers) for food and protein synthesis. Chemical recycling occurs through respiration, the excretion of waste products and, of course, the death of organisms.

> **6. e.** *Students know* a vital part of an ecosystem is the stability of its producers and decomposers.

An ecosystem's producers (plants and photosynthetic microorganisms) and decomposers (fungi and microorganisms) are primarily responsible for the productivity and recycling of organic matter, respectively. Conditions that threaten the stability of producer and decomposer populations in an ecosystem jeopardize the availability of energy and the capability of matter to recycle in the rest of the biological community. To study the interaction between producers and decomposers, students can set up a closed or restricted ecosystem, such as a worm farm, a composting system, a terrarium, or an aquarium.

> **6. f.** *Students know* at each link in a food web some energy is stored in newly made structures but much energy is dissipated into the environment as heat. This dissipation may be represented in an energy pyramid.

The energy pyramid illustrates how stored energy is passed from one organism to another. At every level in a food web, an organism uses energy metabolically to survive and grow, but much is released as heat, usually about 90 percent. At every link in a food web, energy is transferred to the next level, but typically only 10 percent of the energy from the previous level is passed on to the consumer.

> **6. g.*** *Students know* how to distinguish between the accommodation of an individual organism to its environment and the gradual adaptation of a lineage of organisms through genetic change.

Living organisms may adapt to changing environments through nongenetic changes in their structure, metabolism, or behavior or through natural selection of favorable combinations of alleles governing any or all of these processes. Genetic and behavioral adaptations are sometimes difficult to identify or to distinguish without studying the organism over a long time. Physical changes are slow to develop in most organisms, requiring careful measurements over many years. Examining fossil ancestors of an organism may help provide clues for detecting adaptation through genetic change. Genetic change can institute behavioral changes, making it all the more complicated to determine whether a change is solely a behavioral accommodation to environmental change.

Through the use of print and online resources in library-media centers, students can research the effects of encroaching urbanization on undeveloped land and consider the effects on specific species, such as the coyote (not endangered) and the California condor (endangered). Such examples can illustrate how some organisms adapt to their environments through learned changes in behavior, and others are unsuccessful in learning survival skills. Over a long time, organisms can also adapt to changing environments through genetic changes, some of which may include genetically determined changes in behavior. Such changes may be difficult to recognize because a long time must elapse before the changes become evident. Studies of the origins of desert pup fish or blind cave fish may help students understand how gradual genetic changes in an organism lead to adaptations to changes in its habitat.

STANDARD SET 7. Evolution (Population Genetics)

This discussion applies to Standard Sets 7 and 8. Students in grades nine through twelve should be ready to explore and understand the concept of biological evolution from its basis in genetics. The synthesis of genetics, and later of molecular biology, with the Darwin-Wallace theory of natural selection validated the mechanism of evolution and extended its scientific impact. Students need to understand that the same evolutionary mechanisms that have affected the rest of the living world have also affected the human species.

Students need to understand that a theory in science is not merely a hypothesis or a guess, but a unifying explanation of observed phenomena. Charles Darwin's theory of the origin of species by natural selection is such an explanation. Even though biologists continue to test the boundaries of this theory today, their investigations have not found credible evidence to refute the theory. Scientists have also had many opportunities to demonstrate the gradual evolution of populations in the wild and in controlled laboratory settings. As more populations of organisms are studied at the level of DNA sequence and as the fossil record improves, the understanding of species divergence has become clearer.

> **7. The frequency of an allele in a gene pool of a population depends on many factors and may be stable or unstable over time. As a basis for understanding this concept:**
>
> **a.** *Students know* why natural selection acts on the phenotype rather than the genotype of an organism.

Natural selection works directly on the expression or appearance of an inherited trait, the *phenotype,* rather than on the gene combination that produces that trait, the *genotype.* The influence of a dominant allele for a trait over a recessive one in the genotype determines the resulting phenotype on which natural selection acts.

> **7. b.** *Students know* why alleles that are lethal in a homozygous individual may be carried in a heterozygote and thus maintained in a gene pool.

Two types of allele pairings can occur in the genotype: *homozygous* (pairing two of the same alleles, whether dominant, codominant, or recessive) and *heterozygous* (pairing of two different alleles). Recessive lethal alleles (e.g., Tay-Sachs disease) will, by definition, cause the death of only the homozygous recessive individual. Healthy heterozygous individuals will also contribute the masked recessive gene to the population's gene pool, allowing the gene to persist.

> **7. c.** *Students know* new mutations are constantly being generated in a gene pool.

Mutation is an important source of genetic variation within a gene pool. These random changes take the form of additions, deletions, and substitutions of nucleotides and of rearrangements of chromosomes. The effect of many mutations is minor and neutral, being neither favorable nor unfavorable to survival and reproduction. Other mutations may be beneficial or harmful. The important principle is that culling, or selective breeding, cannot eliminate genetic diseases or unwanted traits from a population. The trait constantly reappears in the population in the form of new, spontaneous mutations.

> **7. d.** *Students know* variation within a species increases the likelihood that at least some members of a species will survive under changed environmental conditions.

As environmental factors change, natural selection of adaptive traits must also be realigned. Variation within a species stemming either from mutation or from genetic recombination broadens the opportunity for that species to adapt to change, increasing the probability that at least some members of the species will be suitably adapted to the new conditions. Genetic diversity promotes survival of a species should the environment change significantly, and sameness can mean vulnerability that could lead to extinction.

> **7. e.*** *Students know* the conditions for Hardy-Weinberg equilibrium in a population and why these conditions are not likely to appear in nature.

The principle of Hardy-Weinberg equilibrium, derived in 1908, asserts that the genetic structure of a nonevolving population remains constant over the generations. If mating in a large population occurs randomly without the influence of natural selection, the migration of genes from neighboring populations, or the occurrence of mutations, the frequency of alleles and of genotypes will remain constant over time. Such conditions are so restrictive that they are not likely to occur in nature precisely as predicted, but the Hardy-Weinberg equilibrium equation often gives an excellent approximation for a limited number of generations in sizeable, randomly mating populations. Even though genetic recombination is taken into account, mutations, gene flow between populations, and environmental changes influencing pressures of selection on a population do not cease to occur in the natural world.

> **7. f.*** *Students know* how to solve the Hardy-Weinberg equation to predict the frequency of genotypes in a population, given the frequency of phenotypes.

The Hardy-Weinberg equilibrium equation can be used to calculate the frequency of alleles and genotypes in a population's gene pool. When only two alleles for a trait occur in a population, the letter p is used to represent the frequency of one allele, and the letter q is used to represent the frequency of the other. Students should agree first that the sum of the frequencies of the two alleles is 1, and this equation is written $p + q = 1$. That is, the combined frequencies of the alleles account for all the genes for a given trait.

Students should then consider the possible combinations of alleles in a *diploid organism* (the genome of a diploid organism consists of two copies of each chromosome). An individual could be homozygous for one allele (pp) or homozygous for the other (qq) or heterozygous (either pq or qp). These diploid genotypes will appear at frequencies that are the product of the allele frequencies (e.g., the frequency of a diploid pp individual is p^2, and the frequency of a diploid qq individual is q^2).

The heterozygotes are of two varieties, pq and qp (because the p allele might have been inherited from either parent), but the products of frequency pq and qp are the same. Therefore, the frequency of heterozygotes can simply be expressed as $2pq$. The sum of the frequencies of the homozygous and heterozygous individuals must equal 1, since all individuals have been accounted for. These principles are usually expressed as the equation $p^2 + 2pq + q^2 = 1$. Both equations represent different statements. The first ($p + q = 1$) is an accounting of the two types of alleles in the population, and the second ($p^2 + 2pq + q^2 = 1$) is an accounting of the three distinguishable genotypes.

If the allele frequencies are known (e.g., if $p = 0.1$ and $q = 0.9$) and Hardy-Weinberg equilibrium is assumed, then the frequencies p^2, $2pq$, and q^2 are respectively 0.01, 0.18, and 0.81. That is, 81 percent of individuals would be homozygous qq. If p were a dominant (but nonselective) allele, then $p^2 + 2pq$, or 19 percent of the population, would express the dominant phenotype of the p allele.

The calculation can be used in reverse as well. If Hardy-Weinberg equilibrium conditions exist and 81 percent of the population expresses the qq recessive phenotype, then the allele frequency q is the square root of 0.81, and the rest of the terms can be calculated in a straightforward fashion.

Students can convince themselves of the state of equilibrium by constructing a Punnett Square that assumes random mating. The scenario might be a mass spawning of fish, in which 100,000 eggs and sperm are mixed in a stream and meet with each other randomly to form zygotes. Students can calculate the fraction of p and q type gametes in the stream by thinking through the types of gametes produced by heterozygous and homozygous adult fish. (For this exercise to work, the genotype distribution of adults must agree with Hardy-Weinberg equilibrium.) With the frequencies or numbers of each type of zygote calculated in the cells of a Punnett Square, students will see that equilibrium is preserved. Frequencies of alleles and genotypes, which are the genetic structure of the study population, would remain constant for generations under the premise of Hardy-Weinberg equilibrium.

STANDARD SET 8. Evolution (Speciation)

See the discussion introducing the concept of biological evolution that appears under Standard Set 7, "Evolution (Population Genetics)," on page 237.

8. Evolution is the result of genetic changes that occur in constantly changing environments. As a basis for understanding this concept:

  a. *Students know* how natural selection determines the differential survival of groups of organisms.

Genetic changes can result from gene recombination during gamete formation and from mutations. These events are responsible for variety and diversity within each species. Natural selection favors the organisms that are better suited to survive in a given environment. Those not well suited to the environment may die before they can pass on their traits to the next generation. As the environment changes, selection for adaptive traits is realigned with the change. Traits that were once adaptive may become disadvantageous because of change.

Students can explore the process of natural selection further with an activity based on predator-prey relationships. The main purpose of these activities is to simulate survival in predator or prey species as they struggle to find food or to escape being consumed themselves. The traits of predator and prey individuals can be varied to test their selective fitness in different environmental settings.

An example of natural selection is the effect of industrial "melanism," or darkness of pigmentation, on the peppered moths of Manchester, England. These moths come in two varieties, one darker than the other. Before the industrial revolution, the dark moth was rare; however, during the industrial revolution the light moth seldom appeared. Throughout the industrial revolution, much coal was burned in the region, emitting soot and sulfur dioxide. For reasons not completely understood, the light-colored moth had successfully adapted to the cleaner air conditions that existed in preindustrial times and that exist in the region today.

However, the light-colored moth appears to have lost its survival advantage during times of heavy industrial air pollution. One early explanation is that when soot covered tree bark, light moths became highly visible to predatory birds. Once this change happened, the dark-peppered moth had an inherited survival advantage because it was harder to see against the sooty background. This explanation may not have been the cause, and an alternative one is that the white-peppered moth was more susceptible to the sulfur dioxide emissions of the industrial revolution. In any case, in the evolution of the moth, mutations of the genes produced light and dark moths. Through natural selection the light moth had an adaptive advantage until environmental conditions changed, increasing the population of the dark moths and depleting that of the light moths.

> **8. b.** *Students know* a great diversity of species increases the chance that at least some organisms survive major changes in the environment.

This standard is similar to the previous standard set on diversity within a species but takes student understanding one step further by addressing diversity among and between species. For the same reasons pertinent to those for intraspecies diversity, increased diversity among species increases the chances that some species will adapt to survive future environmental changes.

> **8. c.** *Students know* the effects of genetic drift on the diversity of organisms in a population.

If a small random sample of individuals is separated from a larger population, the gene frequencies in the sample may differ significantly from those in the population as a whole. The shifts in frequency depend only on which individuals fall in the sample (and so are themselves random). Because a random shift in gene frequency is not guaranteed to make the next generation better adapted, the shift—or genetic drift—with respect to the original gene pool is not necessarily an adaptive change. The *bottleneck effect* (i.e., nonselective population reductions due to disasters) and the *founder effect* (i.e., the colonization of a new habitat by a few individuals) describe situations that can lead to genetic drift of small populations.

> **8. d.** *Students know* reproductive or geographic isolation affects speciation.

Events that lead to reproductive isolation of populations of the same species cause new species to appear. Barriers to reproduction that prevent mating between populations are called *prezygotic* (before fertilization) if they involve such factors as the isolation of habitats, a difference in breeding season or mating behavior, or an incompatibility of genitalia or gametes. *Postzygotic* (after fertilization) barriers that prevent the development of viable, fertile hybrids exist because of genetic incompatibility between the populations, hybrid sterility, and hybrid breakdown.

These isolation events can occur within the geographic range of a parent population *(sympatric speciation)* or through the geographic isolation of a small population from its parent population *(allopatric speciation)*. Sympatric speciation is much more common in plants than in animals. Extra sets of chromosomes, or *polypoidy,* that result from mistakes in cell division produce plants still capable of long-term reproduction but animals that are incapable of that process because polypoidy interferes with sex determination and because animals, unlike most plants, are usually of one sex or the other. Allopatric speciation occurs in animal evolution when geographically isolated populations adapt to different environmental conditions. In addition, the rate of allopatric speciation is faster in small populations than in large ones because of greater genetic drift.

> **8. e.** *Students know* how to analyze fossil evidence with regard to biological diversity, episodic speciation, and mass extinction.

Analysis of the fossil record reveals the story of major events in the history of life on earth, sometimes called *macroevolution,* as opposed to the small changes in genes and chromosomes that occur within a single population, or *microevolution.* Explosive radiations of life following mass extinctions are marked by the four eras in the geologic time scale: the Precambrian, Paleozoic, Mesozoic, and Cenozoic. The study of biological diversity from the fossil record is generally limited to the study of the differences among species instead of the differences within a species. Biological diversity within a species is difficult to study because preserved organic material is rare as a source of DNA in fossils.

Episodes of speciation are the most dramatic after the appearance of novel characteristics, such as feathers and wings, or in the aftermath of a mass extinction that has cleared the way for new species to inhabit recently vacated adaptive zones. Extinction is inevitable in a changing world, but examples of mass extinction from the fossil record coincide with rapid global environmental changes. During the formation of the supercontinent Pangaea during the Permian period, most marine invertebrate species disappeared with the loss of their coastal habitats. During the Cretaceous period a climatic shift to cooler temperatures because of diminished solar energy coincided with the extinction of dinosaurs.

> **8. f.*** *Students know* how to use comparative embryology, DNA or protein sequence comparisons, and other independent sources of data to create a branching diagram (cladogram) that shows probable evolutionary relationships.

The area of study that connects biological diversity to *phylogeny,* or the evolutionary history of a species, is called *systematics.* Systematic classification is based on the degree of similarity between species. Thus, comparisons of embryology, anatomy, proteins, and DNA are used to establish the extent of similarities. Embryological studies reveal that *ontogeny,* development of the embryo, provides clues to phylogeny. In contrast to the old assertion that "ontogeny recapitulates phylogeny" (i.e., that it replays the entire evolutionary history of a species), new findings indicate that structures, such as gill pouches, that appear during embryonic development but are less obvious in many adult life forms may establish *homologies* between species (similarities attributable to a common origin). These homologies are evidence of common ancestry. Likewise, homologous anatomical structures, such as the forelimbs of humans, cats, whales, and bats, are also evidence of a common ancestor. Similarity between species can be evaluated at the molecular level by comparing the amino acid sequences of proteins or the nucleotide sequences of DNA strands. DNA-DNA hybridization, restriction mapping, and DNA sequencing are powerful new tools in systematics.

Approaches for using comparison information to classify organisms on the basis of evolutionary history differ greatly. *Cladistics* uses a branching pattern, or *cladogram,* based on shared derived characteristics to map the sequence of evolutionary change. The cladogram is a dichotomous tree that branches to separate those species that share a derived characteristic, such as hair or fur, from those species that lack the characteristic. Each new branch of the cladogram helps to establish a sequence of evolutionary history; however, the extent of divergence between species is unclear from the sequence alone.

Phenetics classifies species entirely on the basis of measurable similarities and differences with no attempt to sort homology from analogy. In recent years phenetic studies have been helped by the use of computer programs to compare species automatically across large numbers of traits. Striking a balance between these two approaches to classification has often involved subjective judgments in the final decision of taxonomic placement. Students can study examples of cladograms and create new ones to understand how a sequence of evolutionary change based on shared derived characteristics is developed.

> **8. g.*** *Students know* how several independent molecular clocks, calibrated against each other and combined with evidence from the fossil record, can help to estimate how long ago various groups of organisms diverged evolutionarily from one other.

Molecular clocks are another tool to establish phylogenetic sequences and the relative dates of phylogenetic branching. Homologous proteins, such as cytochrome c, of different *taxa* (plants and animals classified according to their presumed natural relationships) and the genes that produce those proteins are assumed to evolve at relatively constant rates. On the basis of that assumption, the number of amino acid or nucleotide substitutions provides a record of change proportional to the time between evolutionary branches. The estimates of rate of change derived from these molecular clocks generally agree with parallel data from the fossil record; however, the branching orders and times between branches are more reliably determined by measuring the degree of molecular change than by comparing qualitative features of morphology. When gaps in the fossil record exist, phylogenetic branching dates can be estimated by calibrating molecular change against the timeline determined from the fossil record.

STANDARD SET 9. Physiology (Homeostasis)

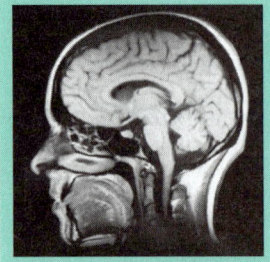

From the individual cell to the total organism, each functioning unit is organized according to *homeostasis,* or how the body and its parts deal with changing demands while maintaining a constant internal environment. In 1859 noted French physiologist Claude Bernard described the difference between the internal environment of the cells and the external environment in which the organism lives. Organisms are shielded from the variations of the external environment by the "constancy of the internal milieu." This "steady state" refers to the dynamic equilibrium achieved by the integrated functioning of all the parts of the organism.

American physiologist Walter Cannon called this phenomenon *homeostasis,* which means "standing still." All organ systems of the human body contribute to homeostasis so that blood and tissue constituents and values stay within a normal range. Students will need supportive review of the major systems of the body and of the organ components of those systems (see Standard Set 2, "Life Sciences," for grade five in Chapter 3 and Standard Set 5, "Structure and Function in Living Systems," for grade seven in Chapter 4). As the prime coordinators of the body's activities, the nervous and endocrine systems must be examined and their interactive roles clearly defined.

9. **As a result of the coordinated structures and functions of organ systems, the internal environment of the human body remains relatively stable (homeostatic) despite changes in the outside environment. As a basis for understanding this concept:**

    a. *Students know* how the complementary activity of major body systems provides cells with oxygen and nutrients and removes toxic waste products such as carbon dioxide.

The digestive system delivers nutrients (e.g., glucose) to the circulatory system. Oxygen molecules move from the air to the alveoli of the lungs and then to the circulatory system. From the circulatory system glucose and oxygen molecules move from the capillaries into the cells of the body where cellular respiration occurs. During cellular respiration these molecules are oxidized into carbon dioxide and water, and energy is trapped in the form of ATP. The gas exchange process is reversed for the removal of carbon dioxide from its higher concentration in the cells to the circulatory system and, finally, to its elimination by exhalation from the lungs.

The concentration of sugar in the blood is monitored, and students should know that sugar can be stored or pulled from reserves (glycogen) in the liver and muscles to maintain a constant blood sugar level. Amino acids contained in proteins can also serve as an energy source, but first the amino acids must be deaminated, or chemically converted, in the liver, producing ammonia (a toxic product), which is converted to water-soluble urea and excreted by the kidneys. Teachers should emphasize that all these chemicals are transported by the circulatory system and the cells. Organs at the final destination direct these chemicals to their exit from the circulatory system.

> **9. b.** *Students know* how the nervous system mediates communication between different parts of the body and the body's interactions with the environment.

An individual becomes aware of the environment through the sense organs and other body receptors (e.g., by allowing for touch, taste, and smell and by collecting information about temperature, light, and sound). The body reflexively responds to external stimuli through a reflex arc (see Standard 9.e in this section). (A *reflex arc* is the pathway along the central nervous system where an impulse must travel to bring about a reflex; e.g., sneezing or coughing.) Students can examine the sense organs, identify other body receptors that make them aware of their environment, and see ways in which the body reflexively responds to an external stimulus through a reflex arc.

Hormones work in conjunction with the nervous system, as shown, for example, in the digestive system, where insulin released from the pancreas into the blood regulates the uptake of glucose by muscle cells. The pituitary master gland produces growth hormone for controlling height. Other pituitary hormones have specialized roles (e.g., follicle-stimulating hormone [FSH] and luteinizing hormone [LH] control the gonads, thyroid-stimulating hormone [TSH] controls the thyroid, and adrenocorticotropic hormone [ACTH] regulates the formation of glucocorticoids by the adrenal cortex). This master gland is itself controlled by the hypothalamus of the brain.

> **9. c.** *Students know* how feedback loops in the nervous and endocrine systems regulate conditions in the body.

Feedback loops are the means through which the nervous system uses the endocrine system to regulate body conditions. The presence or absence of hormones in blood brought to the brain by the circulatory system will trigger an attempt to regulate conditions in the body. To make feedback loops relevant to students, teachers can discuss the hormone leptin, which fat cells produce as they become filled with storage reserves. Leptin is carried by the blood to the brain, where it normally acts to inhibit the appetite center, an example of negative feedback. When fat reserves diminish, the concentration of leptin decreases, a phenomenon that in turn causes the appetite center in the brain to start the hunger stimulus and activate the urge to eat.

> **9. d.** *Students know* the functions of the nervous system and the role of neurons in transmitting electrochemical impulses.

Transmission of nerve impulses involves an electrochemical "action potential" generated by gated ion channels in the membrane that make use of the countervailing gradients of sodium and potassium ions across the membrane. Potassium ion concentration is high inside cells and low outside; sodium ion concentration is the opposite. The sodium and potassium ion concentration gradients are restored by an active transport system, a pump that exchanges sodium and potassium ions across the membrane and uses ATP hydrolysis as a source of free energy. The release of neurotransmitter chemicals from the axon terminal at the synapse may initiate an action potential in an adjacent neuron, propagating the impulse to a new cell.

> **9. e.** *Students know* the roles of sensory neurons, interneurons, and motor neurons in sensation, thought, and response.

The pathways of impulses from dendrite to cell body to axon of sensory neurons, interneurons, and motor neurons link the chains of events that occur in a reflex action. Students should be able to diagram this pathway. Similar paths of neural connections lead to the brain, where the sensations become conscious and conscious actions are initiated in response to external stimuli. Students might also trace the path of the neural connections as the sensation becomes conscious and a response to the external stimulus is initiated. Students should also be able to identify gray and white matter in the central nervous system.

> **9. f.*** *Students know* the individual functions and sites of secretion of digestive enzymes (amylases, proteases, nucleases, lipases), stomach acid, and bile salts.

To bring about digestion, secretions of enzymes are mixed with food (in the mouth and as the food proceeds from the mouth through the stomach and through the small intestines). For example, salivary glands and the pancreas secrete amylase enzymes that change starch into sugar. Stomach acid and gastric enzymes begin the breakdown of protein, a process that intestinal and pancreatic secretions continue.

Lipase enzymes secreted by the pancreas break down fat molecules (which contain three fatty acids) to free fatty acids plus diglycerides (which contain two fatty acids) and monoglycerides (which contain one fatty acid). Bile secreted by the liver furthers the process of digestion, emulsifying fats and facilitating digestion of lipids. Students might diagram the digestive tract, labeling important points of secretion and tracing the pathways from digestion of starches, proteins, and other foods. They can then outline the role of the kidney nephron in the formation of urine and the role of the liver in glucogenesis and glycogenolysis (glucose balance) and in blood detoxification.

9. g.* *Students know* the homeostatic role of the kidneys in the removal of nitrogenous wastes and the role of the liver in blood detoxification and glucose balance.

Microscopic nephrons within the kidney filter out body wastes, regulate water, and stabilize electrolyte levels in blood. The liver removes toxic materials from the blood, stores them, and excretes them into the bile. The liver also regulates blood glucose.

9. h.* *Students know* the cellular and molecular basis of muscle contraction, including the roles of actin, myosin, Ca^{+2}, and ATP.

Controlled by calcium ions and powered by hydrolysis of ATP, actin and myosin filaments in a sarcomere generate movement in stomach muscles. Striated muscle fibers reflect the filamentous makeup and contraction state evidenced by the banding patterns of those fibers. A sketch of the sarcomere can be used to indicate the functions of the actin and myosin filaments and the role of calcium ions and ATP in muscle contraction.

9. i.* *Students know* how hormones (including digestive, reproductive, osmoregulatory) provide internal feedback mechanisms for homeostasis at the cellular level and in whole organisms.

Hormones act as chemical messengers, affecting the activity of neighboring cells or other target organs. Their movement can be traced from their point of origin to the target site. The feedback mechanism works to regulate the activity of hormones and promotes homeostasis.

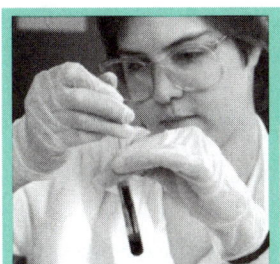

STANDARD SET 10. Physiology (Infection and Immunity)

Some bacteria, parasites, and viruses cause human diseases because they either rob the body of necessary sustenance or secrete toxins that cause injury. The human body has a variety of mechanisms to interfere with or destroy invading pathogens. Besides protection afforded by the skin, one of the most effective means of defending against agents that harm the body is the immune system with its cellular and chemical defenses. Students should develop a clear understanding of the components of the immune system and know how vaccines and antibiotics are used to combat disease. They should also know that acquired immune deficiency syndrome (AIDS) compromises the immune system, causing affected persons to succumb to other AIDS-associated infections that are harmless to people with an intact immune system.

> **10. Organisms have a variety of mechanisms to combat disease. As a basis for understanding the human immune response:**
>
> **a.** *Students know* the role of the skin in providing nonspecific defenses against infection.

The skin serves as a physical barrier to prevent the passage of many disease-causing microorganisms. Cuts and abrasions compromise the skin's ability to act as a barrier. Teachers can use charts and overhead projections to show the dangers and physiologic responses of a break in the skin.

> **10. b.** *Students know* the role of antibodies in the body's response to infection.

Cells produce antibodies to oppose antigens, substances that are foreign to the body. An example of an antigen is a surface protein of a flu virus, a protein with a shape and structure unlike those of any human proteins. The immune system recognizes that the flu virus structure is different and generates proteins called *antibodies* that bind to the flu virus. Antibodies can inactivate pathogens directly or signal immune cells that pathogens are present.

> **10. c.** *Students know* how vaccination protects an individual from infectious diseases.

Several weeks are required before the immune system develops immunity to a new antigen. To overcome this problem, vaccinations safely give the body a look in advance at the foreign structures. Vaccines usually contain either weakened or killed pathogens that are responsible for a specific infectious disease, or they may contain a purified protein or subunit from the pathogen. Although the vaccine does not cause an infectious disease, the antigens in the mixture prompt the body to generate antibodies to oppose the pathogen. When the individual is exposed to the pathogenic agent, perhaps years later, the body still remembers having seen the antigens in the vaccine dose and can respond quickly. Students have been exposed to the practical aspects of immunization through their knowledge of the vaccinations they must receive before they can enter school. They have all experienced getting shots and may have seen their personal vaccination record in which dates and kinds of inoculations are recorded. The review of a typical vaccination record, focusing on the reason for the shots and ways in which they work, may serve as an effective entry to the subject.

Students should review the history of vaccine use. Early literature provides descriptions of vaccine use from pragmatic exposure, but the term *vaccine* is derived from the cowpox exudate that Edward Jenner used during the 1700s to inoculate villagers against the more pathogenic smallpox. Louis Pasteur, noted for his discovery of the rabies treatment, also developed several vaccines. Poliovirus, the cause of infantile paralysis (poliomyelitis), was finally conquered in the 1950s through vaccines that Jonas Salk and Albert B. Sabin refined.

> **10. d.** *Students know* there are important differences between bacteria and viruses with respect to their requirements for growth and replication, the body's primary defenses against bacterial and viral infections, and effective treatments of these infections.

A *virus,* which is the simplest form of a genetic entity, is incapable of metabolic life and reproduction outside the cells of other living organisms. A virus contains genetic material but has no ribosomes. Although some viruses are benign, many harm their host organism by destroying or altering its cell structures. Generally, the body perceives viruses as antigens and produces antibodies to counteract the virus. *Bacteria* are organisms with a full cellular structure. They, too, can be benign or harmful. Harmful bacteria and their toxins are perceived as antigens by the body, which in turn produces antibodies. In some cases infectious diseases may be treated effectively with *antiseptics,* which are chemicals that oxidize or in other ways inactivate the infecting organism. Antiseptics are also useful in decontaminating surfaces with which the body may come in contact (e.g., countertops). *Antibiotics* are effective in treating bacterial infections, sometimes working by destroying or interfering with the growth of bacterial cell walls or the functioning of cell wall physiology or by inhibiting bacterial synthesis of DNA, RNA, or proteins. Antibiotics are ineffective in treating viral infections.

Students might research infections caused by protists (malaria, amoebic dysentery), bacteria (blood poisoning, botulism, food poisoning, tuberculosis), and viruses (rabies, colds, influenza, AIDS). They might also investigate the pathogens currently being discussed in the media and study each infectious organism's requirements for growth and reproduction. Teachers should review the dangers of common bacteria becoming resistant to antibiotics through long-standing overapplication, as shown by the increasing incidence of drug-resistant tuberculosis and other bacteria. Using a commercially available kit, teachers can demonstrate how antibiotics may act generally or specifically against bacteria. Agar plates may be inoculated with different bacteria, and different antibiotic discs may be placed on these plates to create a clear zone in which growth around the antibiotic discs is inhibited.

> **10. e.** *Students know* why an individual with a compromised immune system (for example, a person with AIDS) may be unable to fight off and survive infections by microorganisms that are usually benign.

When an immune system is compromised (e.g., through infection by the human immunodeficiency virus [HIV]), it becomes either unable to recognize a dangerous antigen or incapable of mounting an appropriate defense. This situation happens when the virus infects and destroys key cells in the immune system.

10. f.* *Students know* the roles of phagocytes, B-lymphocytes, and T-lymphocytes in the immune system.

Phagocytes move, amoebalike, through the circulatory system, consuming waste and foreign material, such as aged or damaged blood cells and some infectious bacteria and viruses. Two broad types of lymphocytes (a class of white blood cells) originate in the bone marrow during embryonic life. One type (the B-lymphocyte) matures in the bone marrow and gives rise to antibody-producing plasma cells that are responsible for humoral immunity. Each mature B-lymphocyte can give rise to only a single type of antibody, which itself may recognize only a single foreign antigen. The other type (the T-lymphocyte) matures in the thymus gland during embryogenesis and gives rise to "cytotoxic" (cell killing) and "helper" T-lymphocytes. The cytotoxic T-cells are particularly useful for surveillance of intra-cellular pathogens. Antibodies cannot reach the intracellular pathogen because of the cell membrane, but the infected cell can be identified and killed. Helper T-cells assist in organizing both the humoral and cellular immune responses.

Earth Sciences

By looking outward and deep into space and time, astronomers have discovered a vast and ancient universe. The study of earth sciences helps students find their place in this universe by showing where their unique world fits in with the grand scheme of the cosmos. Students of the earth sciences gain an understanding of the physical and chemical processes that formed Earth and continue to operate on this planet. As students study these science standards, they will also learn more about the geologic factors that help to make California special.

The Sun, a rather ordinary star, provides virtually all the surface energy required for life on Earth. Its energy also drives convection in Earth's atmosphere and oceans, a process that in turn drives global climate conditions and local weather patterns. In addition, heat energy moves slowly below Earth's surface through the planet's interior. Some of this internal heat originated with the formation of the planet, and some is generated by the decay of radioactive nuclides. This geothermal heat slowly escapes to the hydrosphere and atmosphere. The quantity of geothermal heat is tiny compared with the quantity of incoming solar energy. However, over the long term, geothermal heat is responsible for plate tectonic processes—moving continents, building mountains, and causing volcanism and earthquakes.

STANDARD SET 1. Earth's Place in the Universe (Solar System)

Students should previously have studied the star patterns in the night sky and the changes in those patterns with the seasons and lunar cycles. They should also have been introduced to the solar system; and they can be expected to know that the Sun, which is composed primarily of hydrogen and helium, is the center of the solar system. They should also know that the solar system includes Earth and eight other planets, their moons, and a large number of comets and asteroids and that gravitational interaction with the Sun primarily determines the orbits of all these objects. In the eighth grade students should have learned about the composition, relative sizes, positions, and motions of objects in the solar system.

Students should become familiar with evidence that dates Earth at 4.6 billion years old, and they should know that extraterrestrial objects hit the planet occasionally and that such impacts were more frequent in the past. They have also learned that the Moon, planets, and comets shine by reflected light. To study this standard set, students will need to understand electromagnetism and gravity. Students should know and understand the Doppler effect and the inverse square law of light (see Standard 4.f in the physics section of this chapter). Familiarity with the acquisition and analysis of spectral data will also be helpful. The content in this standard set may cause students difficulty in grasping the vastness of geologic time and astronomical distances. Teachers should provide opportunities for students to think about space and time in different scales, from the macroscopic to the microscopic, such as practice in working with relevant numbers and in visualizing the solar system in the appropriate scale.

> **1. Astronomy and planetary exploration reveal the solar system's structure, scale, and change over time. As a basis for understanding this concept:**
>
> **a.** *Students know* how the differences and similarities among the Sun, the terrestrial planets, and the gas planets may have been established during the formation of the solar system.

Students studying this standard will learn how the Sun and planets formed and developed their present characteristics. The solar nebula, a slowly rotating massive cloud of gas and dust, is believed to have contracted under the influence of gravitational forces and eventually formed the Sun, the rocky inner planets, the gaseous outer planets, and the moons, asteroids, and comets. The exact mechanism that caused this event is unknown. The outer planets are condensations of lighter gases that solar winds blew to the outer solar system when the Sun's fusion reaction ignited. Observations supporting this theory are that the orbital planes of the planets are nearly the same and that the planets revolve around the Sun in the same direction.

To comprehend the vast size of the solar system, students will need to understand scale, know the speed of light, and be familiar with units typically used for denoting astronomical distances. For example, Pluto's orbital radius can be expressed as 39.72 AU or 5.96×10^{12} meters or 5.5 light-hours. An astronomical unit (AU) is a unit of length equal to the mean distance of Earth from the Sun, approximately 93 million miles. A light-year, which is approximately 5.88 trillion miles, or 9.46 trillion kilometers, is the distance light can travel through a vacuum in one year. Students can make a scale model to help them visualize the vast distances in the solar system and the relative size of the planets and their orbit around the Sun. Calculator tape may be used to plot these distances to scale.

> **1. b.** *Students know* the evidence from Earth and moon rocks indicates that the solar system was formed from a nebular cloud of dust and gas approximately 4.6 billion years ago.

Since the nineteenth century, geologists, through the use of relative dating techniques, have known that Earth is very old. Relative dating methods, however, are insufficient to identify actual dates for events in the deep past. The discovery of radioactivity provided science with a "clock." Radioactive dating of terrestrial samples, lunar samples, and meteorites indicates that the Earth and Moon system and meteorites are approximately 4.6 billion years old.

The solar system formed from a *nebula,* a cloud of gas and debris. Most of this material consisted of hydrogen and helium created during the big bang, but the material also included heavier elements formed by nucleosynthesis in massive stars that lived and died before the Sun was formed. The death of a star can produce a spectacular explosion called a *supernova,* in which debris rich in heavy elements is ejected into space as stardust. Strong evidence exists that the impact of stardust from a nearby supernova triggered the collapse of the nebula that formed the solar system. The collapse of a nebula leads to heating, an increase in rotation rate, and flattening. From this hot, rapidly spinning nebula emerged the Sun and solid grains of various sizes that later accreted to form objects that evolved through collisions into planets, moons, and meteorites. The nebula from which the Sun and planets formed was composed primarily of hydrogen and helium, and the solar composition reflects this starting mixture. The nebula also contained some heavy elements. As the nebula cooled, condensation of the heavy elements and the loss of volatile elements from the hot, inner nebula led to formation of rocky inner planets. To varying extents, the whole of the solar system was fractionated; but the portion of the solar nebula now occupied by the inner planets was highly fractionated, losing most of its volatile material, while the outer portion (beyond Mars) was less fractionated and is consequently richer in the lighter, more volatile elements.

> **1. c.** *Students know* the evidence from geological studies of Earth and other planets suggests that the early Earth was very different from Earth today.

The prevailing theory is that Earth formed around 4.6 billion years ago by the contraction under gravity of gases and dust grains found in a part of the solar nebula. As Earth accreted, it was heated by the compressing of its material by gravity and by the kinetic energy released when moving bits of debris and even planetoids struck and joined. Eventually, the interior of the planet heated sufficiently for iron, an abundant element in the earth, to melt. Iron's high density caused that element to sink toward the center of Earth. The entire planet differentiated, creating layers with the lower-density materials rising toward the top and the higher-density materials sinking toward the center. The volatile gases were the least dense and were "burped out" to form an atmosphere. The result is Earth's characteristic core, mantle, and crust and its oceans and atmosphere. Overall, Earth has slowly cooled since its formation, although radioactive decay has generated some additional heat.

Evidence from drill core samples and surface exposures of very old rocks reveals that early Earth differed from its present form in the distribution of water, the composition of the atmosphere, and the shapes, sizes, and positions of landmasses. Knowing about the evolution of these systems will help students understand the structure of Earth's lithosphere, hydrosphere, and atmosphere.

The composition of the earliest atmosphere was probably similar to that of present-day volcanic gases, consisting mostly of water vapor, hydrogen, hydrogen chloride, carbon monoxide, carbon dioxide, and nitrogen but lacking in free oxygen. Therefore, no ozone layer existed in the stratosphere to absorb ultraviolet rays, and ultraviolet radiation from the Sun would have kept the surface of the planet sterile. The oldest fossils, which are of anaerobic organisms, indicate that life on

Earth was established sometime before 3.5 billion years ago. Conditions on Earth were suitable for life to originate here, but the possibility that life hitched a ride to this planet on a meteorite cannot be excluded.

The continents have slowly differentiated through the partial melting of rocks, with the lightest portions floating to the top. The absence of atmospheric oxygen permitted substantial quantities of iron (ferrous) to dissolve, and some of this iron later precipitated as iron oxide (ferric oxide or rust) when early photosynthesizers added oxygen to the atmosphere. This precipitation of iron produced "banded iron formations," an important geologic resource for contemporary use. These deposits were formed only during distinct time periods, generally from one to three billion years ago. Subsequently, atmospheric oxygen rose sufficiently to permit multicellular, aerobic organisms to flourish.

> **1. d.** *Students know* the evidence indicating that the planets are much closer to Earth than the stars are.

Observations of planetary motions relative to the seemingly fixed stars indicate that planets are much closer to Earth than are the stars. Direct techniques for measuring distances to planets include radar, which makes use of the Doppler effect. Distances to some nearby stars can be measured by parallax: if a star appears to move slightly with respect to more distant stars as Earth orbits from one side of the Sun to the other, then the angle through which the star appears to move and the diameter of Earth's orbit determine, by the use of simple trigonometry, the distance to the star. For more distant stars and extragalactic objects, indirect methods of estimating distances have to be used, all of which depend on the inverse square law of light. This principle states that the intensity of light observed falls off as the square of the distance from the source.

Student learning activities may include daily observations of the position of the Sun relative to a known horizon, observations of the Moon against the same horizon and also relative to the stars, and observations of planets against the background of stars. Other activities might take advantage of current data on the positions of the planets, computer-based lab exercises, and simulations that incorporate the use of library-media center resources.

> **1. e.** *Students know* the Sun is a typical star and is powered by nuclear reactions, primarily the fusion of hydrogen to form helium.

Comparing the solar spectrum with the spectra of other stars shows that the Sun is a typical star. Analysis of the spectral features of a star provides information on a star's chemical composition and relative abundance of elements. The most abundant element in the Sun is hydrogen. The Sun's enormous energy output is evidence that the Sun is powered by nuclear fusion, the only source of energy that can produce the calculated total luminosity of the Sun over its lifetime. Fusion reactions in the Sun convert hydrogen to helium and to some heavier elements. This conversion is one example of nucleosynthesis, in which the fusion process forms helium and other elements (see Standard 11.c for chemistry in this chapter).

I. f. *Students know* the evidence for the dramatic effects that asteroid impacts have had in shaping the surface of planets and their moons and in mass extinctions of life on Earth.

Impacts of asteroids have created extensive cratering on the Moon, on Mercury, and on other bodies in the solar system. Some craters can also be found on Earth, but most have been destroyed by the active recycling of Earth's planetary surface. Some large impacts have had dramatic effects on Earth and on other planets and their moons. Many believe that the impact of an asteroid produced the unusual iridium-rich layer at the boundary between the rocks of the Cretaceous and the Tertiary periods. This event may have been ultimately responsible for the mass extinction of dinosaurs and many other species 65 million years ago.

Through videos or classroom demonstrations, teachers can introduce simulations of impacts of asteroids. Teachers can model cratering by carefully throwing marbles of different masses (weights) into soft clay or flour at different velocities. Students can observe the patterns of impact and shapes of the craters to help in understanding the physical evidence for impact cratering gathered on Earth and the Moon. Using the mass and velocity of the striking object, students can estimate the energy released from impacts of craters.

I. g.* *Students know* the evidence for the existence of planets orbiting other stars.

Spectral observations and direct imaging of nearby stars show that other stars have planetary systems. In fact, the number of planets that have been discovered to orbit nearby stars is increasing constantly; during 2002 that number exceeded 100. Methods used in these planetary discoveries rely on observing slight oscillations in the star's velocity as revealed by shifts in the frequency of spectral lines. Students can search school and public library collections and appropriate Internet sites for current information about planetary exploration and discoveries of planetary systems.

STANDARD SET 2. Earth's Place in the Universe (Stars, Galaxies, and the Universe)

High school courses in earth sciences will be the first experience for many students in using physical evidence to consider models of stellar life cycles and the history of the universe. Students in earlier grades should have observed the patterns of stars in the sky and learned that the Sun is an average star located in the Milky Way galaxy. Students should also have been introduced to astronomical units (AUs), which measure distances between solar system objects such as Earth and Jupiter. Students should know that distances between stars, and also between galaxies, are measured in parsecs. The *parsec* is the distance at which one astronomical unit subtends one second of arc. This distance is about 3.26 light-years.

The concepts dealt with in this standard set are not a part of students' daily experience. As in the previous standard set, students may need help to internalize the distance and time scales used to describe the universe. In addition, misconceptions derived from outdated hypotheses or from science fiction movies, books, and videos may interfere with developing an understanding of accepted scientific evidence. To promote scientific literacy, school libraries should try to keep their collections up to date. Students can benefit from the significant amount of new data gained from space exploration during the past 20 years.

> **2. Earth-based and space-based astronomy reveal the structure, scale, and changes in stars, galaxies, and the universe over time. As a basis for understanding this concept:**
>
> **a.** *Students know* the solar system is located in an outer edge of the disc-shaped Milky Way galaxy, which spans 100,000 light years.

The solar system is a tiny part of the Milky Way galaxy, which is a vastly larger system held together by gravity and containing gas, dust, and billions of stars. Determining the shape of this galaxy is like reconstructing the shape of a building from the inside. The conception that the Milky Way galaxy is a disc-shaped spiral galaxy with a bulging spherical center of stars is obtained from the location of stars in the galaxy. If viewed under a low-powered telescope from a planet in another galaxy, the Milky Way would look like a fuzzy patch of light. If viewed with more powerful telescopes from that far planet, the Milky Way would look like a typical spiral galaxy. One would need to travel at the speed of light for about 100,000 years to go from one edge of the Milky Way to the galaxy's opposite edge.

> **2. b.** *Students know* galaxies are made of billions of stars and comprise most of the visible mass of the universe.

The large-scale structure of the visible, or luminous, universe consists of stars found by the billions in galaxies. In turn, there are billions of galaxies in the universe separated from each other by great distances and found in groups ranging from a few galaxies to large galaxy clusters with thousands of members. Superclusters are composed of agglomerations of many thousands of galaxy clusters.

Students should know that scientists catalog galaxies and stars according to the coordinates of their positions in the sky, their brightness, and their other physical characteristics. Spectroscopic analysis of the light from distant stars indicates that the same elements that make up nearby stars are present in the Sun, although the percentages of heavy elements may differ.

Matter found in stars makes up most of the mass of the universe's visible matter; that is, matter that emits or reflects light or some other electromagnetic radiation that is detectable on Earth. The presence of otherwise invisible matter can be inferred from the effect of its gravity on visible matter, and the mass of the invisible matter in the universe appears to be even greater than the mass of the visible. To discover what form this invisible (or "dark") matter takes is one of the great goals of astrophysics.

> **2. c.** *Students know* the evidence indicating that all elements with an atomic number greater than that of lithium have been formed by nuclear fusion in stars.

Formation of the elements that compose the universe is called *nucleosynthesis.* Calculations based on nuclear physics suggest that nucleosynthesis occurred through the fusing of light elements to make heavier elements. The composition of distant stars, revealed by their spectra, and the relative abundance of the different elements provide strong evidence that these calculations are correct.

Theoretical models predict that the only elements that should have formed during the big bang are hydrogen, helium, and lithium. All other elements should have formed in the cores of stars through fusion reactions. Fusion requires that one nucleus approach another so closely that they touch and bind together. This process is difficult to accomplish because all nuclei are positively charged and repel their neighbors, creating a barrier that inhibits close approach. However, the barrier can be bypassed if the nuclei have high velocities because of high temperature. Once the process begins, fusion of lightweight nuclei leads to a net release of energy, facilitating further fusion. This mechanism can form elements with nuclei as large as (but no larger than) those of iron, atomic number 26. Temperatures sufficient to initiate fusion are attained in the cores of stars.

In the Sun, and in most stars, hydrogen fusion to form helium is the primary fusion reaction. Elements heavier than carbon are formed only in more massive stars and only during a brief period near the end of their lifetime. A different type of fusion is necessary to form elements heavier than iron. This type can be carried out only by adding neutrons to a preexisting heavy element that forms a "seed." Neutrons are available only during a limited portion of a star's lifetime, particularly during the brief supernova that occurs when a massive star dies.

> **2. d.** *Students know* that stars differ in their life cycles and that visual, radio, and X-ray telescopes may be used to collect data that reveal those differences.

Stars differ in size, color, chemical composition, surface gravity, and temperature, all of which affect the spectrum of the radiation the stars emit and the total energy. It is primarily the electromagnetic radiation emitted from the surface of the Sun and stars that can be detected and studied. Radiation in wavelengths that run from those of X-rays to those of radio waves can be collected by modern telescopes. The data obtained enable astronomers to classify stars, determine their chemical composition, identify the stages of their life cycles, and understand their structures. No one has ever watched a star evolve from birth to death, but astronomers can predict the ultimate fate of a given star by observing many stars at different points in their cycles. The primary characteristics that astronomers use to classify stars are surface temperature and luminosity (the total energy emitted).

> **2. e.*** *Students know* accelerators boost subatomic particles to energy levels that simulate conditions in the stars and in the early history of the universe before stars formed.

Scientists' understanding of processes occurring in stars has been enhanced by experiments in particle physics, nuclear physics, and plasma physics. Particle accelerators create particle velocities great enough for the nuclei of elements to overcome electrostatic repulsion and to approach close enough for nuclear interactions to take place, mimicking stellar nuclear fusion processes. The first accelerator was developed in the 1950s in Berkeley, California. It allowed the energy of protons to be raised high enough to create antimatter particles, thereby making it possible to explore the substructure of what had been considered the most elementary form of matter.

Scientists used the results from these experiments to create models of the processes and conditions under which matter is created. Developed at the turn of the twentieth century, Einstein's special theory of relativity showed that matter and energy are interchangeable. Particle accelerators made it possible to produce, in the laboratory, matter-energy transformations previously possible only in stars. Scientists and engineers continue to look for ways to control and sustain fusion reactions, a potential source for a nearly inexhaustible supply of energy.

> **2. f. *** *Students know* the evidence indicating that the color, brightness, and evolution of a star are determined by a balance between gravitational collapse and nuclear fusion.

A major concept in science is that temperature is a measure of the underlying energy of motion of a system. Furthermore, thermal energy can be radiated away into space as electromagnetic radiation. This process produces the light that Earth receives from the Sun. As the temperature of a star's surface increases, the intensity of radiation produced also increases, and the spectrum of radiation shifts toward a shorter wavelength. Consequently, a blue-white star is hotter than a red star and emits more energy than does a red star of equal size.

A star's surface temperature is a guide to the internal processes occurring within the star. Stars are so hot that they are a form of matter known as a *plasma,* in which atoms move so fast that electrons cannot keep up, leaving the nuclei free as ions. Gravity acts to collapse the ions in the hot plasma. The high density and high temperature of the plasma allow the barrier caused by the mutual repulsion of positive nuclei to be overcome, permitting fusion, or nucleosynthesis, to occur in the stellar core. The energy released from this reaction helps maintain a pressure that resists further compaction by the gravitational force and prevents collapse of the stellar core. The stellar dynamics evolve to a structure that reflects the thermal energy flow from the hot core, where energy is created, to the cooler surface, where it is radiated away to space as starlight. The star will attain an energy balance so that the production of energy by fusion equals the upward heat flow, which in turn equals the energy emitted into space. The size and color of the star reflect the balances needed.

> **2. g.*** *Students know* how the red-shift from distant galaxies and the cosmic background radiation provide evidence for the "big bang" model that suggests that the universe has been expanding for 10 to 20 billion years.

During the 1920s Edwin Hubble observed the *red shift* (the apparent increase in wavelength of emitted radiation) of distant galaxies. The red shift is due to a Doppler effect and indicates that distant galaxies are rapidly receding from ours. He noted that their speed of recession is proportional to their distance and suggested that the universe is expanding. More recent verification from radio waves and other data that a 3K background radiation, or low-level microwave background "noise," exists throughout the universe has led to the acceptance of the big bang model of an expanding universe that is 10 to 20 billion years old. According to this theory, this radiation began as high-energy short-wavelength radiation created by the explosion when the universe was born. As space expanded and the universe cooled down, the wavelengths were essentially stretched out.

A major breakthrough in astrophysics occurred during the 1990s, when scientists at the Lawrence Berkeley National Laboratory in California saw evidence for variation in the intensity of this background radiation. This finding is consistent with the idea that matter in the early universe was already starting to condense in some areas, a necessary first step toward the clumping together that led to the formation of stars and galaxies.

STANDARD SET 3. Dynamic Earth Processes

The earth sciences use concepts, principles, and theories from the physical sciences and mathematics and often draw on facts and information from the biological sciences. To understand Earth's magnetic field and magnetic patterns of the sea floor, students will need to recall, or in some cases learn, the basics of magnetism. To understand circulation in the atmosphere, hydrosphere, and lithosphere, students should know about convection, density and buoyancy, and the Coriolis effect. Earthquake epicenters are located by using geometry. To understand the formation of igneous and sedimentary minerals, students must master concepts related to crystallization and solution chemistry.

Because students in grades nine through twelve may take earth science before they study chemistry or physics, some background information from the physical sciences needs to be introduced in sufficient detail. From standards presented earlier, students should know about plate tectonics as a driving force that shapes Earth's surface. They should know that evidence supporting plate tectonics includes the shape of the continents, the global distribution of fossils and rock types, and the location of earthquakes and volcanoes. They should also understand that plates float on a hot, though mostly solid, slowly convecting mantle. They should be familiar with basic characteristics of volcanoes and earthquakes and the resulting changes in features of Earth's surface from volcanic and earthquake activity.

> **3. Plate tectonics operating over geologic time has changed the patterns of land, sea, and mountains on Earth's surface. As the basis for understanding this concept:**
>
> a. *Students know* features of the ocean floor (magnetic patterns, age, and sea-floor topography) provide evidence of plate tectonics.

Much of the evidence for continental drift came from the seafloor rather than from the continents themselves. The longest topographic feature in the world is the midoceanic ridge system—a chain of volcanoes and rift valleys about 40,000 miles long that rings the planet like the seams of a giant baseball. A portion of this system is the Mid-Atlantic Ridge, which runs parallel to the coasts of Europe and Africa and of North and South America and is located halfway between them. The ridge system is made from the youngest rock on the ocean floor, and the floor gets progressively older, symmetrically, on both sides of the ridge. No portion of the ocean floor is more than about 200 million years old. Sediment is thin on and near the ridge. Sediment found away from the ridge thickens and contains progressively older fossils, a phenomenon that also occurs symmetrically.

Mapping the magnetic field anywhere across the ridge system produces a striking pattern of high and low fields in almost perfect symmetrical stripes. A brilliant piece of scientific detective work inferred that these "zebra stripes" arose because lava had erupted and cooled, locking into the rocks a residual magnetic field whose direction matched that of Earth's field when cooling took place. The magnetic field near the rocks is the sum of the residual field and Earth's present-day field. Near the lavas that cooled during times of normal polarity, the residual field points along Earth's field; therefore, the total field is high. Near the lavas that cooled during times of reversed polarity, the residual field points counter to Earth's field; therefore, the total field is low.

The "stripes" provide strong support for the idea of seafloor spreading because the lava in these stripes can be dated independently and because regions of reversed polarity correspond with times of known geomagnetic field reversals. This theory states that new seafloor is created by volcanic eruptions at the midoceanic ridge and that this erupted material continuously spreads out convectively and opens and creates the ocean basin. At some continental margins deep ocean trenches mark the places where the oldest ocean floor sinks back into the mantle to complete the convective cycle. Continental drift and seafloor spreading form the modern theory of plate tectonics.

> **3. b.** *Students know* the principal structures that form at the three different kinds of plate boundaries.

There are three different types of plate boundaries, classified according to their relative motions: divergent boundaries; convergent boundaries; and transform, or parallel slip, boundaries. *Divergent boundaries* occur where plates are spreading apart. Young divergence is characterized by thin or thinning crust and rift valleys; if divergence goes on long enough, midocean ridges eventually develop, such as the Mid-Atlantic Ridge and the East Pacific Rise.

Convergent boundaries occur where plates are moving toward each other. At a convergent boundary, material that is dense enough, such as oceanic crust, may sink back into the mantle and produce a deep ocean trench. This process is known as *subduction*. The sinking material may partially melt, producing volcanic island arcs, such as the Aleutian Islands and Japan. If the subduction of denser oceanic crust occurs underneath a continent, a volcanic mountain chain, such as the Andes or the Cascades, is formed. When two plates collide and both are too light to subduct, as when one continent crashes into another, the crust is crumpled and uplifted to produce great mountain chains, such as the relatively young Himalayas or the more ancient Appalachians.

The third type of plate boundary, called a *transform,* or *parallel slip, boundary,* comes into existence where two plates move laterally by each other, parallel to the boundary. The San Andreas fault in California is an important example. Marking the boundary between the North American and Pacific plates, the fault runs from the Gulf of California northwest to Mendocino County in northern California.

> **3. c.** *Students know* how to explain the properties of rocks based on the physical and chemical conditions in which they formed, including plate tectonic processes.

Rocks are classified according to their chemical compositions and textures. The composition reflects the chemical constituents available when the rock was formed. The texture is an indication of the conditions of temperature and pressure under which the rock formed. For example, many igneous rocks, which cooled from molten material, have interlocking crystalline textures. Many sedimentary rocks have fragmental textures. Whether formed from cooling magma, created by deposits of sediment grains in varying sizes, or transformed by heat and pressure, each rock possesses identifying properties that reflect its origin.

Plate tectonic processes directly or indirectly control the distribution of different rock types. Subduction, for example, takes rocks from close to the surface and drags them down to depths where they are subjected to increased pressures and temperatures. Tectonic processes also uplift rocks so that they are exposed to lower temperatures and pressures and to the weathering effects of the atmosphere.

> **3. d.** *Students know* why and how earthquakes occur and the scales used to measure their intensity and magnitude.

Most earthquakes are caused by lithospheric plates moving against each other. Earth's brittle crust breaks episodically in a stick-and-slip manner. Plate tectonic stresses build up until enough energy is stored to overcome the frictional forces at plate boundaries. The *magnitude* of an earthquake (e.g., as shown on the Richter scale) is a measure of the amplitude of an earthquake's waves. The magnitude depends on the amount of energy that is stored as elastic strain and then released. Magnitude scales are logarithmic, meaning that each increase of one point on the scale represents a factor of ten increase in wave amplitude and a factor of about thirty increase in energy. An earthquake's intensity (as measured on a modified Mercalli scale) is a subjective, but still valuable, measure of how strong an earthquake felt and how much damage it did at any given location.

> **3. e.** *Students know* there are two kinds of volcanoes: one kind with violent eruptions producing steep slopes and the other kind with voluminous lava flows producing gentle slopes.

The violence of volcanic eruptions is a function of the viscosity of the lava that erupted. All magmas contain dissolved volatiles (or gases) that expand and rise buoyantly as the magma rises to the surface—much like the bubbles in a bottle of soda. Fluid lavas allow gases to bubble away relatively harmlessly, but viscous lavas trap the gases until large pressures build up and the system explodes. Temperature and composition determine the viscosity of magma. Magma at cool temperatures and with a high silica content is very viscous. Rhyolitic and andesitic lavas are examples of lavas with high viscosity. They erupt violently, scattering volcanic fragments and ash widely. Viscous lava, which does not flow very far, builds steep-sided volcanoes. Other lavas, such as basaltic, are relatively fluid and erupt quietly, producing great flows of lava that gradually build gently sloping deposits (called *shield volcanoes*).

> **3. f.*** *Students know* the explanation for the location and properties of volcanoes that are due to hot spots and the explanation for those that are due to subduction.

The melting of silica-rich (granitic) upper-crustal rock produces viscous lavas. The melting of iron-rich (basaltic) lower-crustal, or upper-mantle, rock produces fluid lavas. Upper-crustal rock may melt at subduction zones, and violent volcanic eruptions are common there. Lower-crustal rock may melt at the midocean spreading centers, where quiet, fluid eruptions are common.

Volcanoes may also arise from the activity of mantle plumes, which are long-lived hot spots deep in the mantle. Rock locally melted within the hot spot rises through buoyancy through the crust, sometimes forming volcanoes. As the magma rises, it melts other rocks in its path and incorporates them into the magma. The incorporation of enough upper-crustal rocks, as at the Yellowstone Caldera

Complex at Yellowstone National Park, produces explosive volcanoes. If only lower-crustal rocks are incorporated, as in Hawaii, nonexplosive, gently sloped shield volcanoes form. The Hawaiian Islands are an example of hot spot volcanism, which occurs in chains with the volcanoes systematically aging downward away from the heat source. This type of volcanism is extra evidence supporting the theory of plate tectonics. Volcanoes form when a particular piece of the crust is over the hot spot and then die out as that part of the plate moves off.

STANDARD SET 4. Energy in the Earth System (Solar Energy Enters, Heat Escapes)

Students know that energy is transferred from warmer to cooler objects. They are expected to know that energy is transported by moving material or in heat flow or as waves. They have learned that when fuel is consumed, energy is released as heat, which can be transferred by conduction, convection, or radiation. They have also learned that the Sun is the major source of energy for Earth. They have studied ways in which heat from Earth's interior influences conditions in the atmosphere and oceans and have considered the changes in weather caused by differences in pressure, temperature, air movement, and humidity.

Photosynthesis may have been covered in detail if the students have completed high school biology. Students who have completed high school physics and chemistry will also be better prepared to deal with transfer and absorption of energy. To complete this standard set, students should review the characteristics of the electromagnetic spectrum. Students should also review information presented in the sixth grade science standards related to dynamic Earth processes to increase their awareness of the enormous amount of energy stored in the planet, both as original heat and as a product of radioactive decay. Students should also have studied the mechanisms, primarily mantle convection and some conduction, that bring heat to Earth's surface. Students should know that heat from Earth's interior escapes into the atmosphere through volcanic eruptions, hot spring activity, geysers, and similar means. Although spectacular and energetic, these phenomena are localized and occur over a tiny percentage of Earth's surface. Beyond these readily noticeable losses of interior heat, internal heat disperses into the atmosphere slowly and relatively uniformly across the entire surface of the planet.

> **4. Energy enters the Earth system primarily as solar radiation and eventually escapes as heat. As a basis for understanding this concept:**
>
> a. *Students know* the relative amount of incoming solar energy compared with Earth's internal energy and the energy used by society.

Most of the energy that reaches Earth's surface comes from the Sun as electromagnetic radiation concentrated in infrared, visible, and ultraviolet wavelengths.

The energy available from the Sun's radiation exceeds all other sources of energy available at Earth's surface. There is energy within Earth, some of which is primitive, or original, heat from the planet's formation and some that is generated by the continuing decay of radioactive elements. Over short periods of time, however, only a small amount of that energy reaches Earth's surface. The enormous amount of energy remaining within Earth powers plate tectonics.

Human societies use energy for heating, lighting, transportation, and many other modern conveniences. Most of this energy came to Earth as solar energy. Some has been stored as fossil fuels, plants that stored energy through photosynthesis. Fossil fuels, including oil, natural gas, and coal, provide the majority of energy used by contemporary economies. This energy, which has been stored in crustal rocks during hundreds of millions of years, is ultimately limited. On average a U.S. household consumes energy at the rate of about 1 kilowatt, or 1,000 joules of energy, per second. The Sun delivers almost this much power to every square meter of the illuminated side of Earth. For this reason total energy use by humans is small relative to the total solar energy incident on Earth every day, but harvesting this energy economically poses a challenge to modern engineering.

> **4. b.** *Students know* the fate of incoming solar radiation in terms of reflection, absorption, and photosynthesis.

The fate of incoming solar radiation, which is concentrated in the visible region of the electromagnetic spectrum, is determined by its wavelength. Longer wavelength radiation (e.g., infrared) is absorbed by atmospheric gases. Shorter wavelengths of solar electromagnetic energy, particularly in the visible range, are not absorbed by the atmosphere, except for the absorption of ultraviolet radiation by the ozone layer of the upper atmosphere. Some of the incident visible solar radiation is reflected back into space by clouds, dust, and Earth's surface, and the rest is absorbed.

Plants and other photosynthetic organisms contain chlorophyll that absorbs light in the orange, short-red, blue, and ultraviolet portions of the solar radiation spectrum. The absorption of visible light is less for green and yellow wavelengths, the reflection of which accounts for the color of leaves. The plant uses the absorbed light energy for photosynthesis, in which carbon dioxide and water are converted to sugar, a process that is used to support plant growth and cell metabolism. A by-product of photosynthesis is oxygen. The amount of carbon dioxide in the atmosphere declines slightly during the summer growing season and increases again in the winter. The solar energy stored in plants is the primary energy source for life on Earth.

> **4. c.** *Students know* the different atmospheric gases that absorb the Earth's thermal radiation and the mechanism and significance of the greenhouse effect.

Every object emits electromagnetic radiation that is characteristic of the temperature of the object. This phenomenon is called "blackbody" radiation. For example, an iron bar heated in a fire glows red. At room temperatures the radiation emitted by the bar is in the far infrared region of the electromagnetic spectrum and cannot be seen except with cameras with infrared imaging capability.

The Sun is much hotter than Earth; therefore, energy reaching Earth from the Sun has, on average, much shorter wavelengths than the infrared wavelengths that Earth emits back into space. Energy reaching Earth is mostly in the visible range, and a portion of this energy is absorbed. However, for the planet to achieve energy balance, all the incoming solar energy must be either reflected or reradiated to space. Earth cools itself as the Sun does, by emitting blackbody radiation; but because Earth is cooler than the Sun, Earth's radiation peaks in the infrared instead of in the visible wavelengths.

Certain gases, particularly water vapor, carbon dioxide, methane, and some nitrogen oxide pollutants, transmit visible light but absorb infrared light. These atmospheric constituents, therefore, admit energy from the Sun but inhibit the loss of that energy back into space. This phenomenon is known as the greenhouse effect, and these constituents are called greenhouse gases. Without them Earth would be a colder place in which to live. Human activity, such as the burning of fossil fuels, is increasing the concentration of greenhouse gases in the atmosphere. This buildup can potentially cause a significant increase in global temperatures and affect global and regional weather patterns. Predicting the precise long-term impact is difficult, however, because the influence of cloud cover and other factors is poorly understood.

> **4. d.*** *Students know* the differing greenhouse conditions on Earth, Mars, and Venus; the origins of those conditions; and the climatic consequences of each.

Atmospheric conditions on Earth, Mars, and Venus are different. With a thick atmosphere rich in greenhouse gases, Venus exhibits a much higher planetary surface temperature than does Earth. Mars has a very thin atmosphere depleted in greenhouse gases and therefore has little greenhouse warming. And because Mars lacks oceans and the thin atmosphere does not effectively store heat, the planet experiences large temperature swings: high during the daytime and low at night.

The greenhouse effect is important to Earth's climate because without that effect the planet would be much colder and more like Mars. But if the concentration of absorbing gases is too high, trapping too much heat in the atmosphere, excessive heating could occur on Earth, producing global warming and a climate closer to that of Venus.

The concentration of greenhouse gases, principally that of carbon dioxide, is increasing in Earth's atmosphere, a phenomenon caused primarily by the burning of fossil fuels for electricity and heat. Computer models of the greenhouse effect (a projected buildup of greenhouse gases) predict an increase in average global temperatures. If these models are accurate, the change predicted could have significant consequences on weather patterns and ocean levels. However, Earth's climate system consists of a complex set of positive and negative feedback mechanisms that are not fully understood, and therefore predictions of changes in global temperatures contain some uncertainty.

STANDARD SET 5. Energy in the Earth System (Ocean and Atmospheric Convection)

Students know that the uneven heating of Earth causes air movements and that oceans and the water cycle influence weather. They also know that heat energy is transferred by radiation, conduction, and convection and that radiation from the Sun is responsible for winds and ocean currents, which in turn influence the weather and climate. They should have learned the concept of density and that warm, less-dense fluids rise and cooler, denser fluids sink (see Standard Set 8, "Density and Buoyancy," for grade eight in Chapter 4). Students who have completed courses in chemistry and physics know that water has high heats of crystallization and evaporation and high specific heat (see Standard 7.d for chemistry in this chapter). Others will have to be introduced to these concepts. This knowledge provides a foundation of physical principles for a fuller understanding of energy flow through Earth's system.

> **5. Heating of Earth's surface and atmosphere by the sun drives convection within the atmosphere and oceans, producing winds and ocean currents. As a basis for understanding this concept:**
>
> a. *Students know* how differential heating of Earth results in circulation patterns in the atmosphere and oceans that globally distribute the heat.

The Sun's rays spread unequally across Earth's surface, heating it more at the equator and less at the poles. As heat at the surface transfers to the atmosphere, circulation cells are created. At the equator, for example, hot, moist air rises, expands under lower atmospheric pressure, and cools, causing the air to release its water as precipitation. The air then moves either north or south away from the equator. In its eventual descent the air is compressed by higher atmospheric pressure and warms. Therefore, the air arrives at Earth's surface in a state of low relative humidity. The air then flows back to the equator, completing the cycle. There are three such cycles (or cells) between the equator and the pole. The circulation in these cells regulates the general pattern of rainfall on Earth's surface, with wet climates to be found under ascending air and dry climates under descending air. Therefore, wet climates are generally found at the equator, dry climates in bands at around 30 degrees north and south, wet climates in bands at around 60 degrees, and dry climates again at still higher latitudes.

The same unequal heating of Earth's surface that drives the global atmospheric circulation also causes large thermally driven currents in the oceans. These currents are important in global redistribution of heat. Air currents also distribute heat. Some of the atmospheric heat transport is carried out by exchanging warm and cold air, but water vapor is also a major transport mechanism. Heat is stored in water that evaporates at low latitudes and then is released when the water recondenses (as precipitation) at higher latitudes. For all these reasons combined, the equatorial regions are somewhat cooler, and the poles somewhat warmer, than might otherwise be expected.

Earth's axis is tilted with respect to the plane of its orbit around the Sun. As a result, different amounts of solar energy reach the two hemispheres at different times, thus causing the seasons.

The ocean and atmosphere are a linked system as energy is exchanged between them. Ocean currents rise in part because cool or more saline waters descend, setting circulation patterns in motion. These currents also distribute heat from the equator toward the pole.

> **5. b.** *Students know* the relationship between the rotation of Earth and the circular motions of ocean currents and air in pressure centers.

Earth rotates on an axis, and all flow of fluids on or below the surface appears to be deflected by the Coriolis effect, making right turns in the Northern Hemisphere and left turns in the south. This is a complicated phenomenon to explain to students, but it can be illustrated with a rotatable globe and chalk. Students can hold the globe still and draw a chalk line from the North Pole to the equator and another from the South Pole to the equator. The result will be a part of a great circle. Next the students draw the same line while, at the same time, slowly rotating the globe. A curved line will appear. The faster the globe turns, the more profound the turning of the chalk line. Teachers may find it helpful to compare this effect with centrifugal force, another apparent force arising from an accelerating reference frame. Many good demonstrations of this phenomenon are possible. Teachers can also point out to students that the airflow past a bicycle rider feels the same if the bicycle is still and the air is moving or vice versa. An observer standing on Earth feels that the air is moving, even if the relative motion arises because he or she and Earth are moving through the air.

Combining convective air or water flows with Coriolis turning produces circular currents. For example, when a region, or cell, of lower-pressure (less dense) air exists in the Northern Hemisphere, higher-pressure air tries to flow toward it from all sides by convection. However, the Coriolis effect deflects these flows to the right, leading to a circular airflow, which appears counterclockwise when viewed from above.

> **5. c.** *Students know* the origin and effects of temperature inversions.

Normally, the atmosphere is heated from below by the transfer of energy from Earth's surface. This activity produces *convection,* the transfer of heat by the vertical movements of air masses. However, in certain geographical settings, local sources or sinks for heat can interact with topography to create circumstances in which lower-density warm air, flowing from one direction, is emplaced over higher-density cool air that has come from another direction. This situation, called a *temperature inversion,* effectively stops convection, causing stagnant air. In areas with high population density (or with other sources of pollution) atmospheric pollutants, known as smog, may be trapped by the inversion.

In southern California inverted air occurs normally during the late spring and summer, when the land's temperature is significantly warmer than the ocean's. Air that has been cooled over the ocean flows inland but is stopped by the mountains. Airflow from the deserts, which are at higher elevations, provides warm air that caps this cool marine layer, producing an inversion. This low-elevation, cooler air is held in place by mountains ringing the Los Angeles Basin and is rapidly filled with pollutants.

> **5. d.** *Students know* properties of ocean water, such as temperature and salinity, can be used to explain the layered structure of the oceans, the generation of horizontal and vertical ocean currents, and the geographic distribution of marine organisms.

In low latitudes water is warmed at the surface by the Sun. Differences in the density of water force this water to flow to high latitudes, where it cools as it transfers thermal energy into the atmosphere. Because cooling increases water's density (down to a temperature of 4 degrees Celsius in the case of fresh water and down to the freezing point in the case of sea water), water sinks at high latitudes, flows back toward the equator at depth, and upwells toward the surface as it is warmed by the Sun. This density-driven circulation creates a layered ocean structure at low and midlatitudes, with warm low-density water at the surface and cool high-density water at depth. Salinity also plays a role because rapid evaporation in dry-latitude belts concentrates the salt. Fresh water inflowing from rainfall in wet climatic belts, from rivers, and from melting ice formed at high latitudes decreases salinity.

Because water has a high specific heat, it effectively transports heat from the equator to the poles. Furthermore, the high specific heat helps to buffer Earth's surface against significant daily or seasonal temperature changes. Ice, the solid phase of water, is less dense than the liquid phase and thus floats. (This unique property of water is important to life on Earth.) Icebergs float long distances from their places of origin before they melt and add fresh water to the surface of the ocean.

Water is an excellent solvent for many ions and dissolved gases necessary to sustain marine life. The ocean's chemistry reflects the combined influences of ocean circulation and of marine organisms on biologically active compounds. Water near the surface is oxygenated by photosynthesis, and dissolved nutrients required by phytoplankton are depleted. Zooplankton eat phytoplankton, and the remains of both sink into deeper waters where they decompose. The decomposition enriches deep water in nutrients and depletes it in oxygen, leading to a chemically stratified ocean. Deep water upwelled into the surface zone carries abundant nutrients needed to sustain the growth of phytoplankton. These patterns influence the distribution of marine life because organisms tend to follow and stay within zones that best meet their requirements for survival.

In addition to the density factors that drive ocean circulation, a wind-driven circulation exists in surface waters. These surface and deep currents mix the oceans continuously, particularly at the surface. Ocean currents influence regional climates. For example, the Gulf Stream brings warm water offshore to northwest Europe, warming the climate in such countries as Great Britain. Without these currents the climate would be very different.

To demonstrate the density of currents, teachers can use containers of water heated to different temperatures. Food coloring may be used to dye hot water one color and cold water another. The students observe as the hot water is poured into the cold water and vice versa. To extend the demonstration further, the teacher can add table salt to make different concentrations of salt in same- and different-colored water samples. As the teacher pours saline water into fresh water, and vice versa, students can observe and report what happens.

> **5. e.** *Students know* rain forests and deserts on Earth are distributed in bands at specific latitudes.

Latitudinal bands, or zones, of similar climatic conditions circle Earth. These bands are produced by the large-scale convective air patterns described earlier, known as "Hadley cells." Basically, air rises at the equator and at near 60 degrees north and south latitudes and sinks near 30 degrees north and south latitudes and at both poles. Students will learn these concepts more easily if they understand the ideal gas law and also the notion of relative humidity—that cooler air evaporates less water than does warmer air. If students have not studied these topics, the teacher can explain that sinking air is compressed because of gravity's pull on the overlying air.

Rising air expands and cools, and sinking air is compressed and heated (e.g., compressing air into a bicycle tire warms the air). Because more water can evaporate at warmer temperatures, the air seems drier as it is compressed and heated. Therefore, deserts are common in bands of sinking air and, conversely, high precipitation in zones of rising air supports lush vegetation (e.g., rainforests).

> **5. f.*** *Students know* the interaction of wind patterns, ocean currents, and mountain ranges results in the global pattern of latitudinal bands of rain forests and deserts.

As air is warmed in the tropics, water is evaporated, and the resulting warm, moist air rises and cools. When this moist tropical air cools enough, it becomes saturated and precipitates water as rain. The once warm, moist air is now dryer and cold and heavy. This air is then displaced to the north or south by rising currents of warm, moist air. The cold, dry air begins to descend and is again compressed and heated. At last reaching the ground, at about 30 degrees latitude, the now warm, dry air evaporates water from the ground, producing a desert. A similar pattern is seen farther north and south, where temperate rainforests exist at about 60 degrees latitude, reflecting the rising air in that region. The air sinks at the poles and is warmed somewhat but is still very cold and dry.

Deserts, called *rain shadow deserts,* are also found outside the latitudinal band of deserts. An example is the desert in much of Nevada east of the Sierra Nevada. Warm, moist winds blowing off the Pacific Ocean rise up over the Sierra Nevada, cooling and dropping rain on the forested westward side of the mountains. East of the mountains the air, which is dry, drops down to lower elevations, heats up, and evaporates surface water, producing a desert.

Global weather and atmospheric circulation maps from the weather bureau are helpful for studying this process. Such maps can be downloaded from appropriate Internet sites. Students may search an atlas for maps that show the distribution of deserts and rain forests and compare those maps with global weather maps. Students can plot atmospheric and oceanic currents on a world map and identify regions that are warm and wet and those that are cold and dry.

> **5. g.*** *Students know* features of the ENSO (El Niño southern oscillation) cycle in terms of sea-surface and air temperature variations across the Pacific and some climatic results of this cycle.

The *El Niño* southern oscillation (ENSO) cycle refers to the observed relationships between periodic changes in the patterns of temperature and air pressure of the equatorial surface of the Pacific Ocean and overlying air masses. These relationships change on a time scale of several years and correlate with characteristic variability in global weather climate. Data on sea surface temperatures gathered during several decades can be compared with other records for weather and related topics to see patterns develop (e.g., temperatures near the coast of southern California can be compared with rainfall totals in various places in the world or with various agricultural indicators).

STANDARD SET 6. Energy in the Earth System (Climate and Weather)

This standard set is designed to help students focus on the various factors that produce climate and weather. Since the study of the water (hydrologic) cycle is fundamental to understanding weather, teachers should review that cycle during the study of Standard Set 6. In standard sets taught previously in the lower grade levels, weather was introduced, as a phenomenon, followed by a discussion of the procedures in which weather is observed, measured, and described. Subsequently, weather maps were introduced, and students should have learned to read and interpret topographic maps. The Investigation and Experimentation standards for grades six and seven also called for students to construct scale models and make predictions from accumulated evidence. Teachers should review the concept of pressure with the students.

> **6. Climate is the long-term average of a region's weather and depends on many factors. As a basis for understanding this concept:**
>
> **a.** *Students know* weather (in the short run) and climate (in the long run) involve the transfer of energy into and out of the atmosphere.

Unequal transmission and absorption of solar energy cause differences in air temperature and therefore differences in pressure; winds are generated as a result. Solar-influenced evaporation and precipitation of water determine the humidity of the atmosphere. Evaporation and precipitation also transfer energy between the atmosphere and oceans because energy is absorbed when water evaporates and is released when water condenses. Climate is the long-term average of weather. According to an old saying, "Climate is what you expect, and weather is what you get."

> **6. b.** *Students know* the effects on climate of latitude, elevation, topography, and proximity to large bodies of water and cold or warm ocean currents.

Previous earth science standards covered how and why the locations of rainforests and deserts depend on latitude. But other variables can modify the climate in a particular region. For example, since air expands and cools when it rises, expected temperatures at high elevations are considerably lower than they are at sea level or below. Mountains also affect local climate because of the rain-shadow effect, described in Standard Set 5, "Energy in the Earth System," in this section, and the direction of prevailing winds. The Indian monsoon cycle and the smaller-scale Santa Ana winds are other examples of how mountains may influence weather and climate.

The proximity of land to large bodies of water can also strongly influence climate. Large-scale warm and cold oceanic currents (e.g., the cold Japanese current off the coast of California and the warm Gulf Stream off the East Coast of the United States) exert regional controls on the climate of adjoining landmasses. Even more important, water has a very high specific heat, which causes water to remain within a relatively narrow temperature range between day and night and from season to season. Because of this phenomenon, regions near bodies of water have a tempered climate generally cooler than inland regions during hot weather and warmer than inland regions during cold weather.

> **6. c.** *Students know* how Earth's climate has changed over time, corresponding to changes in Earth's geography, atmospheric composition, and other factors, such as solar radiation and plate movement.

Because Earth is dynamic, particularly with regard to long-term changes in the distribution of continents caused by plate tectonic movements, the planet's climate has changed over time. Some geologic eras were much warmer than the present Cenozoic era. At other times much of the land was covered in giant ice sheets.

Astronomical factors that vary significantly only over millenia and such factors as changes in the tilt of Earth's axis of rotation and changes in the shape of Earth's orbit also influence climate. The configuration of continental landmasses affects ocean currents. Climate is affected, episodically, by volcanic eruptions and impacts of meteorites that inject dust into the atmosphere. Dust and volcanic ash reduce the amount of energy penetrating the atmosphere, thereby changing atmospheric circulation, rainfall patterns, and Earth's surface temperatures.

Variations in life in general, and human activity in particular, affect the amounts of carbon dioxide and other gases that enter the atmosphere. The effect of carbon dioxide and other greenhouse gases is discussed in Standard 4.d in this section.

> **6. d.*** *Students know* how computer models are used to predict the effects of the increase in greenhouse gases on climate for the planet as a whole and for specific regions.

Scientists now know enough about what controls complex climatic variations to construct computer-generated models on global and regional scales. Such models can now make useful predictions about the consequences of greenhouse gases, including the potential for changes in global and regional mean temperatures. Computer-generated weather models have been improved and broadened sufficiently to be useful in exploring long-term changes in weather that border on climatic predictors. Specific models have been constructed to predict effects of ozone depletion and an increase in greenhouse gases.

Students can download current and historical data on weather from the Internet and use this information to explore whether a correlation exists between data on weather and on greenhouse gas production. Students' conclusions may be compared with results from computer-generated greenhouse models and interpretations published in scientific papers or posted on the Internet. However, students should be advised to expect contrary opinions because interpretations of the same climatic data can vary widely.

STANDARD SET 7. Biogeochemical Cycles

Students who complete high school biology/life sciences before they take earth sciences will already have learned about biogeochemical cycles. Through standards presented in lower grade levels, other students should have been exposed to life cycles, food chains, and the movement of chemical elements through biological and physical systems. Students should also have studied chemical changes in organisms and should know that through photosynthesis solar energy is used to create the molecules needed by plants. In this standard set students will learn that within the biogeochemical cycles, matter is transferred between organisms through food webs or chains. Matter can also be transferred from these cycles into physical environments where the cycling elements are held in reservoirs. Matter can be transferred back into biological cycles through physical processes, such as volcanic eruptions and products of the rock cycle, particularly those from weathering.

> **7. Each element on Earth moves among reservoirs, which exist in the solid earth, in oceans, in the atmosphere, and within and among organisms as part of biogeochemical cycles. As a basis for understanding this concept:**
>
> **a.** *Students know* the carbon cycle of photosynthesis and respiration and the nitrogen cycle.

Carbon and nitrogen move through biogeochemical cycles. The recycling of these components in the environment is crucial to the maintenance of life. Through photosynthesis, carbon is incorporated into the biosphere from the atmosphere. It is then released back into the atmosphere through respiration. Carbon dioxide in the atmosphere is dissolved and stored in the ocean as carbonate and bicarbonate ions, which organisms take in to make their shells. When these organisms die, their shells rain down to the ocean floor, where they may be dissolved if the water is not saturated in carbonate. Otherwise, the shells are deposited on the ocean floor and become incorporated into the sediment, eventually turning into a bed of carbonate rock, such as limestone.

Uplifted limestone may dissolve in acidic rain to return carbon to the atmosphere as carbon dioxide, sending calcium ions back into the ocean where they will precipitate with carbon dioxide to form new carbonate material. Carbonate rocks may also be subducted, heated to high temperatures, and decomposed, returning carbon to the atmosphere as volcanic carbon dioxide gas. Carbon is also stored in the solid earth as graphite, methane gas, petroleum, or coal.

Nitrogen, another element important to life, also cycles through the biosphere and environment. Nitrogen gas makes up most of the atmosphere, but elemental nitrogen is relatively inert, and multicellular plants and animals cannot use it directly. Nitrogen must be "fixed," or converted to ammonia, by specialized bacteria. Other bacteria change ammonia to nitrite and then to nitrate, which plants can use as a nutrient. Eventually, decomposer bacteria return nitrogen to the atmosphere by reversing this process.

> **7. b.** *Students know* the global carbon cycle: the different physical and chemical forms of carbon in the atmosphere, oceans, biomass, fossil fuels, and the movement of carbon among these reservoirs.

The global carbon cycle extends across physical and biological Earth systems. Carbon is held temporarily in a number of reservoirs, such as in biomass, the atmosphere, oceans, and in fossil fuels. Carbon appears primarily as carbon dioxide in the atmosphere. In oceans carbon takes the form of dissolved carbon dioxide and of bicarbonate and carbonate ions. In the biosphere carbon takes the form of sugar and of many other organic molecules in living organisms. Some movement of carbon between reservoirs takes place through biological means, such as respiration and photosynthesis, or through physical means, such as those related to plate tectonics and the geologic cycle. Carbon fixed into the biosphere and then transformed into coal, oil, and gas deposits within the solid earth has in recent years been returning to the atmosphere through the burning of fossil fuels to generate energy. This release of carbon has increased the concentration of carbon dioxide in the atmosphere. Carbon dioxide is a primary greenhouse gas, and its concentration in the atmosphere is tied to climatic conditions.

> **7. c.** *Students know* the movement of matter among reservoirs is driven by Earth's internal and external sources of energy.

The energy to move carbon from one reservoir to another originates either from solar energy or as heat from Earth's interior. For example, the energy that plants use for photosynthesis comes directly from the Sun, and the heat that drives subduction comes from the solid earth.

> **7. d.*** *Students know* the relative residence times and flow characteristics of carbon in and out of its different reservoirs.

Carbon moves at different rates from one reservoir to another, measured by its residence time in any particular reservoir. For example, carbon may move quickly from the biomass to the atmosphere and back because its residence time in organisms is relatively short and the processes of photosynthesis and respiration are relatively fast. Carbon may move very slowly from a coal deposit or a fossil fuel to the atmosphere because its residence time in the coal bed is long and oxidation of coal by weathering processes is relatively slow.

STANDARD SET 8. Structure and Composition of the Atmosphere

Students have little direct background on the structure and composition of the atmosphere beyond what they have learned from Standard Set 7, "Biogeochemical Cycles." If they have taken high school biology/life sciences before studying earth sciences, they will know how organisms exert chemical influences on the air around them through photosynthesis and respiration.

> **8.** Life has changed Earth's atmosphere, and changes in the atmosphere affect conditions for life. As a basis for understanding this concept:
>
>   **a.** *Students know* the thermal structure and chemical composition of the atmosphere.

The atmosphere is a mixture of gases: nitrogen (78 percent), oxygen (21 percent), argon (1 percent), and trace gases, such as water vapor and carbon dioxide. Gravity pulls air toward Earth, and the atmosphere gradually becomes less dense as elevation increases. The atmosphere is classified into four layers according to the temperature gradient. The temperature decreases with altitude in the troposphere, the first layer; then similarly increases in the stratosphere, the second layer; decreases in the mesosphere, the third layer; and increases in the thermosphere (ionosphere), the fourth layer.

The *troposphere,* the layer in which weather occurs, supports life on Earth. The *stratosphere* is less dense than the troposphere but has a similar composition except that this second layer is nearly devoid of water. The other difference is that solar radiation ionizes atoms in the stratosphere and dissociates oxygen to form ozone, O_3. This process is important to life on Earth because ozone absorbs harmful ultraviolet radiation that would otherwise cause health problems. Air in the *mesosphere* has very low density and is ionized by solar radiation. The *thermosphere,* the outermost layer of the atmosphere, is almost devoid of air and receives the direct rays of the Sun. The thermosphere provides a good illustration of the difference between temperature and heat. Temperature is high there because the little heat absorbed is distributed among very few molecules, keeping the average energy of each molecule high.

> **8. b.** *Students know* how the composition of the Earth's atmosphere has evolved over geologic time and know the effect of outgassing, the variations of carbon dioxide concentration, and the origin of atmospheric oxygen.

During the early history of the solar system, strong solar winds drove the primordial atmosphere away. This atmosphere was then replaced by a combination of gases released from Earth (outgassing), mostly through volcanic action, and by bombardment of materials from comets and asteroids. Chemical reactions through time, in the presence of water, changed the atmosphere's original methane and ammonia into nitrogen, hydrogen, and carbon dioxide. Lightweight hydrogen escaped, leaving a predominance of nitrogen. As life capable of photosynthesis developed on Earth, carbon dioxide was taken up by plants, and oxygen was released. The present balance of gases in the atmosphere was achieved at least 600 million years ago. Small but important variations in the amount of carbon dioxide in the atmosphere have occurred naturally since then. Significant increases have been measured in modern times and attributed in large part to human activities, such as the burning of fossil fuels.

> **8. c.** *Students know* the location of the ozone layer in the upper atmosphere, its role in absorbing ultraviolet radiation, and the way in which this layer varies both naturally and in response to human activities.

The ozone layer in the stratosphere is formed when high-energy solar radiation interacts with diatomic oxygen molecules (O_2) to produce ozone, a triatomic oxygen molecule (O_3). By absorbing ultraviolet radiation, the ozone eventually converts back to diatomic oxygen. This absorption of ultraviolet radiation in the stratosphere reduces radiation levels at Earth's surface and mitigates harmful effects on plants and animals. The formation and destruction of ozone creates an equilibrium concentration of ozone in the stratosphere. A reduction in stratospheric ozone near the poles has been detected, believed to be caused by the release of chlorofluorocarbons (CFCs), such as those used as working fluids in air conditioners. The halogens in these CFCs interfere with the formation of ozone by acting as catalysts—substances that modify the rate of a reaction without being consumed in the process. As catalysts, a few molecules of CFCs can help to eliminate hundreds of ozone molecules in the stratosphere. While ozone is beneficial in the stratosphere, it is also a manufactured photochemical pollutant in the lower atmosphere. Students should be taught the importance of reducing the level of ozone in the troposphere and of maintaining the concentration of that gas in the stratosphere.

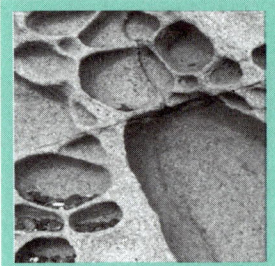

STANDARD SET 9. California Geology

Students should already know that mountains, faults, and volcanoes in California result from plate tectonic activity and that flowing surface water is the most important agent in shaping the California landscape. The topics in this standard set can be covered as a separate unit or as a part of a unit included in other topics addressed by the standards. A specific discussion of California earthquakes can be introduced in the teaching of Standard Set 3, "Dynamic Earth Processes," in this section.

9. **The geology of California underlies the state's wealth of natural resources as well as its natural hazards. As a basis for understanding this concept:**

   a. *Students know* the resources of major economic importance in California and their relation to California's geology.

Many of the important natural resources of California are related to geology. The Central Valley is a major agricultural area and a source of oil and natural gas because of deposition of sediments in the valley, which was created by faulting that occurred simultaneously as the Sierra Nevada was elevated tectonically. California's valuable ore deposits (e.g., gold) came into existence during the formation of large igneous intrusions, when molten igneous rock was injected into older rocks. Geothermal energy resources are related to mountain building and to plate tectonic spreading, or rifting, of the continent.

9. b. *Students know* the principal natural hazards in different California regions and the geologic basis of those hazards.

California is subject to a variety of natural hazards. Active fault zones generate earthquakes, such as those of the San Andreas fault system. Uplifted areas with weak underlying rocks and sediments are prone to landslides, and the California Cascade mountains contain both active and dormant volcanoes. The erosion of coastal cliffs is expected, caused in part by the energy of waves eroding them at their bases. When earthquakes occur along the Pacific Rim, seismic sea waves, or tsunamis, may be generated.

> **9. c.** *Students know* the importance of water to society, the origins of California's fresh water, and the relationship between supply and need.

Water is especially important in California because its economy is based on agriculture and industry, both of which require large quantities of water. California is blessed with an abundance of fresh water, which is supplied by precipitation and collected from the melting of the snowpack in watersheds located in the Sierra Nevada and in other mountain ranges. This process ensures a slow runoff of water following the winter rains and snowfall. But the water is not distributed evenly. Northern California receives most of the rain and snowfall, and southern California is arid to semiarid. The natural distribution of water is adjusted through engineered projects that transport water in canals from the northern to the southern part of the state.

> **9. d.*** *Students know* how to analyze published geologic hazard maps of California and know how to use the map's information to identify evidence of geologic events of the past and predict geologic changes in the future.

Students who learn to read and analyze published geological hazard maps will be able to make better personal decisions about the safety of business and residential locations. They will also be able to make intelligent voting decisions relative to public land use and remediation of hazards.

A wealth of information pertaining to these content standards for earth science is readily available, much of it on the Internet. County governments have agencies that dispense information about resources and hazards, often related to issuing permits and collecting taxes. The California Division of Mines and Geology is an excellent state-level resource. Federal agencies that supply useful information about California resources and hazards are the U.S. Geological Survey, the Federal Emergency Management Agency, and the U.S. Army Corps of Engineers.

Investigation and Experimentation

Teachers must convey the skills and knowledge that students need to perform investigations and experiments, the foundation of scientific knowledge. The Investigation and Experimentation standards allow students to make concrete associations between science and the study of nature and to provide many opportunities to take measurements and use basic mathematics. In the sequence for grades nine through twelve, teachers implement these standards in the context of physics, chemistry, biology/life sciences, and earth sciences.

Investigations and experiments engage scientists, catalyzing their highest levels of creativity and producing their most satisfying rewards. The possibility of discovery or of adding new scientific knowledge in the form of facts, concepts, principles, or theories offers a great sense of accomplishment and wonder. Investigation and experimentation can be just as engaging to high school students as they study science. Although students may not discover knowledge new to the scientific community, they may find pleasure in discovering something new to themselves or in seeing the content from their science text illuminated through demonstrations of the concepts. Accordingly, they can experience the pride of creating experimental protocols and realize the joy of discovery and learning.

Teachers need to know and teach the details of the scientific processes addressed by the Investigation and Experimentation standards. To be valid, an experiment needs controls that minimize sources of error and that provide reproducible results. Teachers should select standards-based, well-tested experiments and demonstrations instead of unguided or disorganized "expeditions." Taught effectively, science courses may be engaging for high school students. Some principles are best pretaught explicitly through direct instruction, then demonstrated with a hands-on activity that reinforces the teaching. Students may easily discover other principles by themselves, and teachers should not rob them of that pleasure. The teacher must be certain that every investigative activity reinforces content and sound thinking.

1. **Scientific progress is made by asking meaningful questions and conducting careful investigations. As a basis for understanding this concept and addressing the content of the other four strands, students should develop their own questions and perform investigations. Students will:**

    a. Select and use appropriate tools and technology (such as computer-linked probes, spreadsheets, and graphing calculators) to perform tests, collect data, analyze relationships, and display data.

    b. Identify and communicate sources of unavoidable experimental error.

c. Identify possible reasons for inconsistent results, such as sources of error or uncontrolled conditions.

d. Formulate explanations by using logic and evidence.

e. Solve scientific problems by using quadratic equations and simple trigonometric, exponential, and logarithmic functions.

f. Distinguish between hypothesis and theory as scientific terms.

g. Recognize the usefulness and limitations of models and theories as scientific representations of reality.

h. Read and interpret topographic and geologic maps.

i. Analyze the locations, sequences, or time intervals that are characteristic of natural phenomena (e.g., relative ages of rocks, locations of planets over time, and succession of species in an ecosystem).

j. Recognize the issues of statistical variability and the need for controlled tests.

k. Recognize the cumulative nature of scientific evidence.

l. Analyze situations and solve problems that require combining and applying concepts from more than one area of science.

m. Investigate a science-based societal issue by researching the literature, analyzing data, and communicating the findings. Examples of issues include irradiation of food, cloning of animals by somatic cell nuclear transfer, choice of energy sources, and land and water use decisions in California.

n. Know that when an observation does not agree with an accepted scientific theory, the observation is sometimes mistaken or fraudulent (e.g., the Piltdown Man fossil or unidentified flying objects) and that the theory is sometimes wrong (e.g., the Ptolemaic model of the movement of the Sun, Moon, and planets).

Assessment of Student Learning

Assessment of Student Learning

Good science assessment answers the critical question: Have the students mastered the content? The answer is used to guide curriculum and instruction. Science assessments need to cover all the grade-level standards. Although some standards will be readily taught and assessed, others will require intensive and extensive instruction accompanied by matching assessments. Assessments also need to reflect the balance among and emphasis given to the earth, life, and physical sciences and the Investigation and Experimentation strands.

California's Standardized Testing and Reporting (STAR) program currently includes standards-based assessments at the high school level (grades nine through eleven) for students enrolled in physics, chemistry, biology/life sciences, earth sciences, and integrated science courses. In the near future a standards-based science assessment for students in grade five (covering the science standards for grades four and five) will be developed and implemented. The STAR program also includes a nationally norm-referenced science test for all students in grades nine through eleven. STAR program results are important to educators at the state and local levels, parents or guardians, and students.

This chapter supports two key principles of assessment in science:

- All students need and are entitled to assessment at appropriate times during the school year on the science standards they are learning.
- All teachers need and are entitled to receive assessment results in a timely manner that empower them to assist students meaningfully in mastering the content.

These principles apply equally to students in kindergarten through grade eight—all of whom receive science instruction regularly—and high school students who are enrolled in standards-based, laboratory science courses. The challenge to educators and to instructional materials publishers is to fashion valid and reliable assessments that are appropriately integrated in a program of efficient and effective science instruction.

Purposes of Assessment

Assessment may serve many purposes, but three are of particular importance:

- Entry-level assessment measures the extent of students' existing knowledge and skill, helping teachers to determine whether review is needed for certain content and whether some students are ready for greater challenges.

- Progress-monitoring assessment measures the extent to which students have mastered (or are mastering) science content sufficiently to proceed in the logical sequence of instruction.
- Summative assessment measures the extent to which students have mastered science content, understand the content well, and are able to apply the knowledge meaningfully.

Entry-Level Assessment

Entry-level assessment measures student mastery of preceding sets of content standards that serve as prerequisite building blocks for the content forthcoming. Typically, the prerequisite knowledge was learned in preceding years or in separate courses. Entry-level assessment also helps the teacher determine what (if anything) in the planned course of instruction has already been mastered by students. Basic psychometric principles must be followed for an entry-level assessment to be used reliably and effectively in comparing the performance of students in a classroom (or school) or to establish a baseline by which to measure growth. The principles are as follows:

- The assessment must be administered under the same conditions to all students.
- The assessment must be administered using the same directions to all students.
- The assessment must be scored in scaled increments small enough to detect growth.

Progress-Monitoring Assessment

Assessment to monitor students' progress is critical in standards-based instruction. Adopted science instructional programs have regular assessments embedded in them. Assessment is the means by which teachers can continually adjust instruction so that all students make progress toward content mastery. Every student in need of extra help or a different instructional approach to master content must receive this type of assessment quickly. Similarly, students who are ready to move on must be enabled to do so and not be required to spend time in unnecessary review.

Teachers need to look regularly for indicators of content mastery in homework and classroom learning activities. Short, objective assessments (e.g., weekly or even daily quizzes) may also prove useful. In addition to this regular monitoring of achievement, more formal assessment needs to be conducted at least every six weeks of instruction. In all cases, progress-monitoring assessments must:

- Use uniform administration procedures and tasks.
- Document student performance.
- Reflect current lessons.
- Help teachers make solid instructional decisions and adjustments based on student performance.
- Indicate when direct interventions are needed for students who are struggling to master content standards.

Summative Assessment

Summative assessment is typically conducted at the end of a chapter, unit, or school year. It measures students' ability to apply the science knowledge and skill they have acquired. Summative assessment requires students to demonstrate understanding of the facts, concepts, principles, and theories in the science standards.

Science Assessment Strategies

Teachers confront a complex array of responsibilities, including the selection and implementation of effective instructional methods and assessment strategies. To help ease teachers' workload in those areas, it is important for publishers of science instructional materials to incorporate multiple measures of assessment that may be used at various points in the school year (or course). The measures need to:

- Reveal the student's knowledge and skill in science and the ability to apply that knowledge and skill as a foundation for future learning.
- Document the student's progress (or lack of progress) toward mastering the content standards.
- Provide information useful to planning and modifying future instruction in ways that will help all students master (or exceed) the content standards.
- Help identify and reinforce effective instructional practices.

The types of assessments are as follows:

Multiple-Choice, Short-Answer, and Essay Responses

Multiple-choice and short-answer tests are familiar, basic assessment instruments. They are particularly useful in covering a number of topics quickly and, if appropriately prepared, provide valid and reliable evaluations of student achievement.

Essay responses are useful for exploring in greater depth students' ability to apply the facts, concepts, principles, and theories learned in science. They typically take more time than multiple-choice or short-answer tests in relation to the number of topics covered, and they are also more complex to grade validly and reliably. They are usually graded with a rubric created in advance. In evaluating essay responses in the area of science, teachers need to be careful to differentiate between students' actual lack of knowledge (or misunderstanding) of science content and limited writing ability.

Investigation and Experimentation Assessments

Investigations and experiments typically involve either (1) a clear-cut question and application of an experimental procedure or protocol; or (2) analysis of a problem and student selection of an appropriate procedure or protocol to use in solving it. In either approach relevant data may be collected or analyzed by students, and the results and conclusions communicated orally or in writing.

As with assessment of essay responses, teachers must be careful in assessing students who conduct investigations and experiments. Teachers may have to distinguish between students' actual lack of knowledge (or misunderstanding) of science content and physical or linguistic limitations.

Reasonable measures may need to be taken to accommodate individual students, but the rigor of the learning and assessment challenge must be equivalent for all students.

STAR Program Results

The State Board of Education has adopted five performance standards for reporting student achievement on the STAR program's standards-based tests in science. The performance standards (levels) are Advanced, Proficient, Basic, Below Basic, and Far Below Basic. Over time, student achievement in science in relationship to these performance standards will be incorporated in California's Academic Performance Index (API), a measure of overall student achievement in the California public schools. Through the API, growth targets are established for individual schools each year, and a system of rewards and interventions has been created based on meeting (or failing to meet) those growth targets.

The STAR program includes four standards-based, subject-specific tests for high school science: physics, chemistry, biology/life sciences, and earth sciences. Standards-based tests have also been created for integrated science courses at the high school level by using various groupings of the same items that appear on the subject-specific tests. All the tests include assessment in the Investigation and Experimentation standards, using items that reflect the science content. These tests are taken only by students enrolled in the respective courses.

Summary

All students need and are entitled to assessment at appropriate times during the school year on the science standards they are learning. Teachers need to receive timely assessment results that inform instruction and curriculum to assist students meaningfully in mastering the content. Three principal types of assessment are important to science instruction: entry-level assessment, progress-monitoring assessment, and summative assessment. Assessments need to reflect the richness of the science standards and, in the elementary and middle grades, to cover all the strands (life, physical, and earth sciences) and the Investigation and Experimentation standards.

7

Universal Access to the Science Curriculum

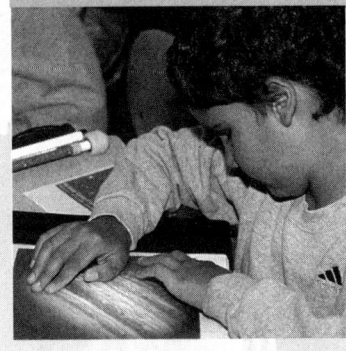

Universal Access to the Science Curriculum

Science education is intended for all students. Academic instruction must be designed so that each student has the opportunity to master the science standards that provide systematic and coherent access to this challenging subject. Toward that end, this chapter discusses all students' needs to develop basic skills and academic language. It addresses the special needs of English learners and students with disabilities in their quest to master the standards, including those in the Investigation and Experimentation strand. The chapter also emphasizes that advanced learners need to be given meaningful assignments that extend the depth and breadth of their understanding of the standards.

Science and Basic Skills Development

The acquisition of scientific knowledge and ideas requires foundational and fundamental skills in English–language arts and mathematics. Students pursuing the science content reflected in the standards and described in this framework need to master the grade-level standards in English–language arts and mathematics. The standards for both science and English–language arts call for students to read and comprehend informational text and write grammatically correct expository essays. Because of the strong relationship between mastery of basic skills in reading, writing, and mathematics and mastery of science content, teachers must reinforce application of those skills in science instruction. Teachers must also ensure that students learn to use scientific terminology correctly and develop academic language (as discussed more fully below).

Academic Language Development

Studying science involves acquiring a new vocabulary and learning that some familiar words may have different meanings in science. This aspect of scientific literacy needs to be taught explicitly in order to minimize misconceptions that might otherwise arise from word usage in differing environments. For example, the terms *control* and *theory* have different definitions in common use compared with scientific use. Students will also begin to acquire new terms that have Latin and Greek roots, prefixes, and suffixes. An understanding of root words and affixes will not only improve vocabulary but also increase students' ability to comprehend words they have not encountered before. For example, students will come to know that *biology* is a combination of *bio-*, derived from the Greek word for life, and *-logy* (also rooted in Greek), meaning study.

English Learners

Support for English learners may consist of the preteaching of essential elements of scientific vocabulary. Instruction for English learners in the academic language of science is critical and must be specifically designed, planned, and taught. It includes direct instruction and experiences for students in English phonology, morphology, syntax, and semantics, and it must support students as they move toward proficiency in the academic language of science. The Investigation and Experimentation strand provides an additional opportunity for teachers to reinforce English learners' understanding of the academic language of science.

In the *Reading/Language Arts Framework* English learners are classified as follows:[1]

- Students in kindergarten through grade two
- Students in grade three through twelve who are literate in their primary language
- Students in grades three through twelve who have limited prior academic experience or literacy in their primary language

The biggest challenge in the teaching of science is addressing the needs of English learners in the latter two groups. Students in grades three through twelve who have strong literacy skills in their primary language can be expected to transfer many of those skills to English and to progress rapidly in learning the academic language of science. Students in grades three through twelve with limited prior schooling will require intensive support in learning the academic language of science.

Advanced Learners

Ensuring mastery of the science standards through challenging and enriched instruction is the goal for advanced learners. Students who readily understand the basic underpinnings of the standards pursue a richer understanding of standards-based science content. Advanced learners in kindergarten through grade eight must be encouraged to extend their knowledge of science through the enrichment opportunities included in state-adopted instructional materials. Enrichment lessons have high levels of standards-based science content proportionate to the amount of time that the lessons take. For example, advanced learners are encouraged to explore the history of a scientific concept or a complex method of experimentation. Enrichment projects need to be designed so that the student does most of the work in the classroom.

Students with Disabilities

Students with disabilities are provided with access to all the content standards through a rich and supported program that uses instructional materials and strategies that best meet their needs. A student's 504 accommodation plan[2] or individualized education program (IEP)[3] often includes suggestions for a variety of techniques to ensure that the student has full access to a program designed to provide him or her with mastery of the science standards, including those in the Investigation and Experimentation strand. Teachers must familiarize themselves with each student's 504 accommodation plan or IEP to help the student in achieving mastery of the science standards.

There are numerous ways in which a teacher can implement accommodations in science instruction. Disabilities vary widely, and accommodations must be tailored to the student's individual and unique needs. Some accommodations help ensure safety while the students participate in investigation and experimentation activities. Examples of some simple safety accommodations are as follows:

- Use of tape with a textured surface to help visually impaired students locate buttons and knobs
- Insulation of exposed hot pipes to protect a student who lacks sensation in the lower extremities and who would likely be in a wheelchair
- Benches at an appropriate height for wheelchair access
- Facilities for emergency showers and eyewashes accessible to wheelchair users

Labels may be printed in larger fonts for low-vision students. Because of the incidence of red-green color-blindness in some male students, instructions and safety notices are not to be red-green color coded, and chemical indicators must not be of the type that change from red to green. Details of laboratory instructions and protocols need to be written for students with hearing impairments or with auditory processing disorders. The following measures may be taken to help all children, especially those with sequencing disabilities or attention deficit disorder: printed instructions detailing each step of a laboratory exercise, checklists to indicate whether each step has been accomplished, and color coding information.

Educators may visit the following Web sites to obtain resources for understanding and addressing the needs of students with disabilities:

- "California Special Education Programs: A Composite of Laws Database," *Education Code,* Part 30, Other Related Laws, and *California Code of Regulations, Title 5* at *http://www.cde.ca.gov/sp/se/ds/*

Notes

1. *Reading/Language Arts Framework for California Public Schools, Kindergarten Through Grade Twelve.* Sacramento: California Department of Education, 1999, p. 233.

2. A Section 504 accommodation plan is a document typically produced by school districts in compliance with the requirements of Section 504 of the federal Rehabilitation Act of 1973. The plan specifies agreed-on services and accommodations for a student who, as the result of an evaluation, is determined to have a "physical or mental impairment [that] substantially limits one or more major life activities." In contrast to the federal Individuals with Disabilities Education Act (IDEA), Section 504 allows a wide range of information to be contained in a plan: (1) the nature of the disability; (2) the basis for determining the disability; (3) the educational impact of the disability; (4) necessary accommodations; and (5) the least restrictive environment in which the student may be placed.

3. An IEP is a written, comprehensive statement of the educational needs of a child with a disability and the specially designed instruction and related services to be employed to meet those needs. An IEP is developed (and periodically reviewed and revised) by a team of individuals, including the parent(s) or guardian(s), knowledgeable about the child's disability. The IEP complies with the requirements of the IDEA and covers such items as (1) the child's present level of performance in relation to the curriculum; (2) measurable annual goals related to involvement and progress in the curriculum; (3) specialized programs (or program modifications) and services to be provided; (4) participation with nondisabled children in regular classes and activities; and (5) accommodation and modification in assessments.

Professional Development

Professional Development

Students of science deserve to be taught by teachers who have both the scientific knowledge and teaching ability to provide skillful instruction. The professional development of teachers is important to raise the quality of science instruction and ensure that science instruction is aligned with the content standards that are the basis of the state assessment system. Professional development programs serve many different types of teachers, but all such programs must strive for improved student achievement as the primary objective. To that end programs must be focused on science instruction at each teacher's specific grade level or content strand (e.g., physics, chemistry, biology/life sciences, earth sciences) and the standards associated with that grade level or strand.

Teachers' collegiate backgrounds vary. Therefore, the professional development needs of an elementary school teacher who may not possess a baccalaureate degree in science will certainly differ from the needs of a single-subject high school chemistry teacher who may have a graduate degree in chemistry. Although some teachers may need to be briefed only on the changes in science that have occurred since their postsecondary-level study, others may lack even basic knowledge of science. Programs must be designed for both types of teachers and provide them with the breadth and the depth of knowledge that are required to support successful standards-based science teaching. As explained in the report of the Glenn Commission, "High-quality teaching requires that teachers have a deep knowledge of subject matter. For this there is no substitute."[1] The standards are the best organizing device for professional development programs and need to be at the center of any planning effort.

For kindergarten through grade eight, the State Board of Education adopts instructional materials after a careful review by panels of expert scientists and teachers. Professional development programs that work with elementary and middle school teachers need to provide them with specific training in the use of the state-adopted instructional materials that have been locally selected. State-adopted instructional materials reflect the best practices for instruction aligned with the content standards and provide:

- Comprehensive coverage of the science content in the standards
- Sequential organization that allows teachers to convey the science content efficiently and effectively

- Strategies for assessment of students
- Information and ideas that address the needs of special student populations
- Information on instructional planning and support

These materials also provide curricular units with investigations and experiments that have clear procedures and explanations of the underlying concepts behind the state standards. The programs have been reviewed to ensure that they support understanding of the standards and that activities demonstrate scientific principles, produce meaningful data, and can be safely and inexpensively conducted.

The State Board of Education does not adopt instructional materials for grades nine through twelve, but local educational agencies may use many of the State Board-approved criteria for evaluating materials at the high school level in physics, chemistry, biology/life sciences, and earth sciences. For example, high school texts need to provide accurate and up-to-date science content and use scientific vocabulary correctly. Standards-based laboratory and field activities need to build investigative skills and judgment, logical thinking, and understanding of scientific principles. The instructional program needs to help teachers evaluate the progress of students toward measurable goals and ensure that students master the content standards. Professional development programs that serve teachers in grades nine through twelve should include examples of outstanding instructional materials that are aligned with the *Science Content Standards*.[2]

Teachers with single-subject credentials in science may have baccalaureate degrees in the subject or subjects they are teaching, may have completed a substantial number of collegiate units in the subjects, may have qualified by successful passage of challenge examination(s), or may have any combination of these qualifications. Currently, the minimum requirements by the Commission on Teacher Credentialing for single-subject credentials in science include a breadth of course work in biology, chemistry, geoscience, and physics in addition to a depth of course work that is focused on one of those areas. Additional information about state credentialing for teaching may be obtained at the Web site of the commission <*http://www.ctc.ca.gov*>. Where possible, science teachers need to be encouraged to work with scientists in industry and postsecondary institutions. Toward this end many companies and collegiate science departments are broadening their outreach to schools and teachers.

What Is Professional Development?

Professional development means "a planned, collaborative, educational process of continual improvement for teachers"[3] that helps them to develop the following proficiencies:

- Enhance their capacity to help students master the standards, using State Board of Education-adopted instructional materials that have been selected locally for students in kindergarten through grade eight or are aligned with the standards (grades nine through twelve).[4]
- Deepen their knowledge of the subject(s) they are teaching.

- Sharpen their teaching skills in the classroom.
- Keep up with developments in their fields.
- Increase their ability to monitor students' work, so they can provide constructive feedback to students and appropriately redirect their own teaching.

Professional development programs should not expect teachers to develop their own curriculum units or use "hand-me-down" units that have been informally produced. Those units may not have been adequately reviewed for accuracy by content experts, or they may inadvertently include activities that are unsafe. Teaching is a challenging job that needs to be skillfully performed by professionals. Similarly, curriculum development requires a special type of expertise. The standards-based accountability system in California is a new experience for many teachers, and ensuring that every child receives a standards-based education is a challenge. Teachers need to have outstanding programs developed so that they can be assured that all the content material is covered comprehensively and in the appropriate sequence.

Who Should Teach the Teachers?

The ultimate goal of professional development is to improve students' academic performance. To that end it is essential that the faculty of a professional development program be experts themselves in the science content called for in the standards. In designing a professional development program, organizers need to seek the help of collegiate faculty in academic science departments and the professional scientists in industry. Although many academic scientists are busy professionals, they are also committed to serving the schools in their communities and may even have children in public schools. Whenever possible, scientists need to be more than just visiting speakers. Professional development programs must be logical and coherent in organization, and collegiate-level scientists can play a key role in their successful design.

Individuals who are not experts in science should not be called upon to teach teachers. Just as students deserve to have competent instruction and standards-aligned curricular materials, so too do their teachers deserve to have the very best program of development. Nonscientists can play significant and important supportive roles in a professional development program and should be used in ways that add to the success of the teachers.

When Is a Program Aligned with the Science Content Standards?

Professional development programs need to focus on the content that teachers are called on to teach. It is important that teachers know the background underlying the standards and know how the content is applied in more advanced study. Teachers must be knowledgeable about a wide range of examples that illustrate the standards they are helping students to master and the models that can be used in the presentation of those standards. Experimental activities used in the process must provide a clear demonstration of the content being studied and be built on a solid foundation of prior knowledge.

Elementary and middle school teachers need to be trained specifically in the grade-level standards applicable to their classes. A fifth-grade teacher, for example, would receive minimal benefit from a program designed for second-grade teachers. Similarly, single-subject teachers in middle and high schools need to be developed as experts in their teaching fields. An understanding of the standards will provide teachers with knowledge of what has been taught to their students in previous grades and what will be taught to their students in future grades, but this concern is secondary to the current needs of their students. Professional development must always be highly focused on the grade level (or subject specialty) of the teacher, not be a holistic study of the standards for many grades (or content strands), a course that lacks necessary depth.

When Is a Professional Development Program Deemed Successful?

The academic achievement of students must be the main indicator of success in professional development programs. An effective program of professional development is one in which adequate attention is paid to classroom follow-up. Program organizers need to track the extent to which teachers apply what they have learned in the classroom and observe how the professional development lessons are implemented.

How Will Tomorrow's Science Teachers Be Developed?

Today's teachers need to encourage students (at the elementary, middle, and high school levels) who are interested in science to pursue both collegiate study and careers in teaching. Moreover, students now completing baccalaureate degrees in science need to be directed toward teaching, and transitions to the field of teaching need to be facilitated for individuals who are contemplating midcareer changes and who have professional experience in science. Only through this multifaceted approach can the demands for science teachers be met for forthcoming generations.

Notes

1. *Before It's Too Late: A Report to the Nation from the National Commission on Mathematics and Science Teaching for the 21st Century.* Washington, D.C.: U.S. Department of Education, 2000.
2. *Science Content Standards for California Public Schools, Kindergarten Through Grade Twelve.* Sacramento: California Department of Education, 2002.
3. *Before It's Too Late.*
4. Ibid.

9

Criteria for Evaluating Instructional Materials in Science, Kindergarten Through Grade Eight

 Criteria for Evaluating Instructional Materials in Science, Kindergarten Through Grade Eight

Instructional materials are adopted by the state for the purpose of helping teachers present the content set forth in the *Science Content Standards for California Public Schools* (referred to in this document as the "California Science Standards"). To accomplish that purpose, this document provides the criteria for evaluating the alignment of the instructional materials with the California Science Standards, as defined in *Education Code* Section 60010. These criteria will govern the evaluation of instructional materials for kindergarten through grade eight (K–8) that are submitted for adoption, beginning with the 2006 Adoption of Science Instructional Materials, and will be helpful to publishers in developing their submission.

The California Science Standards are challenging. In the initial years of implementing the 2003 *Science Framework for California Public Schools* (referred to in this document as the "California Science Framework"), a major goal of most local educational agencies across the state is to facilitate the transition from what many students have traditionally been taught in science to the rigorous content presented in the California Science Standards. Instructional materials play a central role in facilitating that transition. Students should have the opportunity to learn science by direct instruction, by reading textbooks and supplemental materials, by solving standards-based problems, and by doing laboratory investigations and experiments.

The State Board of Education (State Board) will adopt science programs that provide effective learning materials for all students—those students who have mastered most of the content taught in the earlier grades and those who have not—and that specifically address the needs of teachers who instruct a diverse student population. Some teachers may not have specialized in science and may not have an extensive background in science; others may hold supplemental authorizations in life or physical sciences or may have had extensive training in science content and pedagogy. The publishers shall develop and submit programs that offer the flexibility to meet the diverse needs of students and teachers with varying science backgrounds.

These criteria, in keeping with the California Science Framework, do not specify a single pedagogical approach, although the framework incorporates certain commonsense pedagogical features. The State Board encourages publishers to select research-based pedagogical approaches that comprehensively cover the rigorous California Science Standards, reflect the California Science Framework, make judicious use of instructional time, present science in interesting and engaging ways, and otherwise give teachers the resources they need to teach science effectively.

The criteria are organized into five categories:

1. **Science Content/Alignment with Standards:** The content as specified in the California Science Standards and presented in accord with the guidance provided in the California Science Framework
2. **Program Organization:** The sequence and organization of the science program that provide structure to what students should learn each year
3. **Assessment:** The strategies presented in the instructional materials for measuring what students know and are able to do
4. **Universal Access:** The resources and strategies that address the needs of special student populations, including students with disabilities, students whose achievement is either significantly below or above that typical of their class or grade level, and students with special needs related to English language proficiency
5. **Instructional Planning and Support:** The instructional planning and support information and materials, typically including a separate edition specially designed for use by the teacher, that enable the teacher to implement the science program effectively

In kindergarten through grade five, the California Science Standards are organized by grade level in three content strands: physical sciences, life sciences, and earth sciences. The standards for grades six through eight provide for a specific content focus in each year: earth sciences in grade six, life sciences in grade seven, and physical sciences in grade eight. Investigation and Experimentation standards are also provided at each grade level (K–8) and must be taught in the context of these content strands.

In grades nine through twelve, the California Science Standards are organized by discipline. A set of Investigation and Experimentation standards common to all the disciplines is also presented. Most high schools provide the grade nine through grade twelve science curriculum in discipline-specific courses, and some either exclusively provide integrated science courses that combine the various disciplines or provide integrated courses in addition to discipline-specific courses. To allow local educational agencies and teachers flexibility in presenting the material, the standards do not identify a particular discipline with a particular grade. Moreover, the standards do not specify a particular organization of the content of each discipline, although the California Science Framework suggests the logical sequencing of content in some places. Instructional materials may group related standards and address them simultaneously for purposes of coherence and utility.

Submissions that fail to meet Category 1, the Science Content/Alignment with Standards criteria, will not be considered satisfactory for adoption. Categories 2 through 5 will be considered as a whole, each submission passing or failing these criteria as a group. However, every submission will be expected to have strengths in each of categories 2 through 5 to be worthy of adoption.

Category 1: Science Content/Alignment with Standards

Science instructional materials must support the teaching and learning of the California Science Standards in accord with the guidance provided in the California Science Framework. To be considered suitable for adoption, instructional materials must provide:

1. Content that is scientifically accurate.
2. Comprehensive teaching of all California Science Standards at the intended grade level(s) as discussed and prioritized in the California Science Framework, chapters 3 and 4. The only standards that may be referenced are the California Science Standards. There should be no reference to national standards or benchmarks or to any standards other than the California Science Standards.
3. Multiple exposures to the California Science Standards (introductory, reinforcing, and summative), leading to student mastery of each standard through sustained effort.
4. A checklist of California Science Standards in the teacher edition, with page number citations or other references that demonstrate multiple points of student exposure, and a reasonable and judicious allotment of instructional time for learning the content of each standard. Extraneous lessons or topics that are not directly focused on the standards are minimal, certainly composing no more than 10 percent of the science instructional time.
5. A table of evidence in the teacher edition, demonstrating that the California Science Standards can be comprehensively taught from the submitted materials with hands-on activities composing at least 20 to 25 percent of the science instructional program. Hands-on activities must be cohesive, be connected, and build on each other to lead students to a comprehensive understanding of the California Science Standards.
6. Investigations and experiments that are integral to and supportive of the grade-appropriate physical, life, and earth sciences standards so that investigative and experimental skills are learned in the context of those content standards. The instructional materials must include clear procedures and explanations, in the teacher and student materials, of the science content embedded in hands-on activities.
7. Evidence in the teacher edition that each hands-on activity directly covers one or more of the standards in the California Science Standards (in the grade-appropriate physical, life, or earth sciences strands); demonstrates scientific concepts, principles, and theories outlined in the California Science Framework; and produces scientifically meaningful data in practice. All hands-on activities must be safe and age appropriate.
8. Explicit instruction in science vocabulary that emphasizes the meanings of roots, prefixes, and suffixes and the usage and meaning of common words in a scientific context.
9. Extensive, grade-level-appropriate reading and writing of expository text and practice in the use of mathematics, aligned with the *Reading/Language Arts*

Framework for California Public Schools and the *Mathematics Framework for California Public Schools,* respectively.

10. Examples, when directly supportive of the California Science Standards, of the historical development of science and its impact on technology and society. The contributions of minority persons, particularly those individuals who are recognized as prominent in their respective fields, should be included and discussed when it is historically accurate to do so.

11. Examples, when directly supportive of the California Science Standards, of the principles of environmental science, such as conservation of natural resources and pollution prevention. These examples should give direct attention to the responsibilities of all people to create and maintain a healthy environment and to use resources wisely.

Category 2: Program Organization

The sequence and organization of the science program provide structure to what students should learn each year and allow teachers to convey the science content efficiently and effectively. The program content is organized and presented in a manner consistent with the guidance provided in the California Science Framework. To be considered suitable for adoption, instructional materials must provide:

1. A logical and coherent structure that facilitates efficient and effective teaching and learning within a lesson, unit, and year.
2. Specific instructional objectives that are identified and sequenced so that prerequisite knowledge is introduced before more advanced content.
3. Clearly stated student outcomes and goals that are measurable and are based on standards.
4. Materials and assessments that include a cumulative or spiraled review of skills.
5. A program organization that provides the option of preparing for or pre-teaching the science content embedded in any hands-on activities.
6. A program organization that supports various lengths of instructional time and helps make efficient use of small blocks of time (that may be available during the instructional day) in kindergarten through grade three.
7. An overview of the content in each lesson or instructional unit that outlines the scientific concepts and skills to be developed. Topical headings need to reflect the framework and standards and clearly indicate the content that follows.
8. Support materials that are an integral part of the instructional program. These may include video and audio materials, software, and student workbooks.
9. Tables of contents, indexes, glossaries, content summaries, and assessment guides that are designed to help teachers, parents/guardians, and students.
10. For grades four through eight, explicit statements of the relevant grade-level standards in both the teacher and student editions.

Category 3: Assessment

Instructional materials should contain strategies and tools for continually measuring student achievement, following the guidance provided in Chapter 6 of the California Science Framework. To be considered suitable for adoption, instructional materials must provide:

1. Strategies or instruments teachers can use to determine students' entry-level skills and knowledge and methods of using that information to guide instruction
2. Multiple measures of the individual student's progress at regular intervals and at strategic points of instruction, such as lesson, chapter, and unit tests or laboratory reports
3. Suggestions on how to use assessment data to guide decisions about instructional practices and to help teachers determine the effectiveness of their instruction
4. Guiding questions for monitoring students' comprehension
5. Answer keys for all workbooks and other related student resources

Category 4: Universal Access

The instructional materials must provide resources and strategies to enable the effective teaching of students with special needs, allowing them full access to the rigorous academic content specified in the California Science Standards, in accordance with the guidance set forth in Chapter 7 of the California Science Framework. The resources and strategies must support compliance with applicable state and federal requirements for providing instruction to diverse populations and students with special needs and should be consistent with any applicable policies of the State Board toward that end. To be considered suitable for adoption, instructional materials must provide:

1. Suggestions, based on current and confirmed research, for strategies to adapt the curriculum and the instruction to meet students' identified special needs
2. Strategies to help students who are below grade level in science learning, including more explicit explanations of the science content, to accelerate their knowledge to grade level
3. Teacher and student editions that include suggestions or reading materials for advanced learners who need an enriched or accelerated program or more complex assignments
4. Suggestions to help teachers pre-teach and reinforce science vocabulary and concepts with English learners
5. Resources that provide specific help to meet the needs of students whose reading, writing, listening, and speaking skills are below grade level (in relation to the *English–Language Arts Content Standards for California Public Schools* and the *Reading/Language Arts Framework for California Public Schools*) and help to ensure that these students know, understand, and use appropriate academic language in science

6. Evidence of adherence to the Design Principles for Perceptual Alternatives, Design Principles for Cognitive Alternatives, and Design Principles for Means of Expression, as detailed below, to allow access for all students:

   Design Principles for Perceptual Alternatives

   - Provide all student text in digital format, consistent with federal copyright law, so that it can easily be transcribed, reproduced, modified, and distributed in braille, large print (only if the publisher does not offer such an edition), recordings, American Sign Language videos, or other specialized accessible media for use by pupils with visual disabilities or other disabilities that prevent the use of standard materials.

   - Provide written captions or written descriptions in digital format for the audio portions of visual instructional materials, such as videotapes (for those students who are deaf or hard of hearing).

   - Provide educationally relevant descriptions of the images, graphic devices, or pictorial information included in the materials that are essential to the teaching of key concepts. (When important information is presented solely in graphic or pictorial form, it limits access for students who are blind or who have low vision. Digital images with verbal descriptions provide access for those individuals and also provide flexibility for instructional emphasis, clarity, and direction.)

   Design Principles for Cognitive Alternatives

   - Use "considerate text" design principles, including the following techniques and practices:
     — Adequate titles for each selection
     — Introductory subheadings for chapter sections
     — Introductory paragraphs for new chapters and sections
     — Concluding or summary paragraphs, where appropriate
     — Complete paragraphs, including clear topic sentences, relevant support for the topic, and transitional words and expressions (e.g., *furthermore, similarly*)
     — Effective use of typographical aids, such as boldface print, italics
     — Adequate, relevant visual aids connected to the text, such as illustrations, photos, graphs, charts, maps
     — Manageable, not overwhelming, visual and print stimuli
     — Identification and highlighting of important terms
     — List of reading objectives or focus questions at the beginning of each selection
     — List of follow-up comprehension and application questions

   - Provide optional information or activities to enhance students' background knowledge. (Some students face barriers because they lack the necessary background knowledge. Pretesting before an activity will alert teachers to the need for advanced preparation. Instructional materials may include optional supports for background knowledge, to be used by students who need them.)

- Provide cognitive supports for content and activities, including the following items:
  — Assessments to determine background knowledge
  — Summaries of those key concepts from the standards that the content addresses
  — Scaffolds for learning and generalization
  — Opportunities to build fluency through practice

Design Principles for Means of Expression

- Explain in the teacher edition that there are various ways for students with special needs to use the materials and demonstrate their competence, and suggest modifications that teachers might use to allow students to do so. For example, for students who have dyslexia (or difficulties physically forming letters, writing legibly, or spelling words), appropriate modifications of means of expression might be (but are not limited to) students' use of computers to complete pencil-and-paper tasks, including the use of on-screen scanning keyboards, enlarged keyboards, word prediction, and spellcheckers.
- Provide support materials that will give students opportunities to develop oral and written expression.

Category 5: Instructional Planning and Support

Instructional materials must contain a clear road map for teachers to follow when planning instruction. To be considered suitable for adoption, instructional materials must provide:

1. A teacher edition that includes ample and useful annotations and suggestions on how to present the content in the student edition and in the ancillary materials.
2. A checklist of program lessons in the teacher edition, with cross-references to the standards covered, and details regarding the instructional time necessary for all instruction and hands-on activities.
3. Lesson plans, including suggestions for organizing resources in the classroom and ideas for pacing lessons.
4. Blackline masters that are accessible in print and in digitized formats and are easily reproduced. Dark areas are to be minimized to conserve toner.
5. Prioritization of critical components of lessons. Learning objectives and instruction are explicit, and the relationship of lessons to standards or skills within standards is explicit.

6. Clear, grade-appropriate explanations of science concepts, principles, and theories that are presented in a form that teachers can easily adapt for classroom use.
7. Lists of necessary equipment and materials for any hands-on activities, guidance on obtaining those materials inexpensively, and explicit instructions for organizing and safely conducting the instruction.
8. Strategies to address and correct common student errors and misconceptions.
9. Suggestions for how to adapt each hands-on activity provided to other methods of teaching, including teacher modeling, teacher demonstration, direct instruction, or reading, as specified in the California Science Framework.
10. Charts of time and cost of staff development services available for preparing teachers to fully implement the science program.
11. Technical support and suggestions for appropriate use of audiovisual, multimedia, and information technology resources associated with a unit.
12. Strategies for informing parents and guardians about the science program and suggestions for how they can help to support student achievement.
13. Teacher editions containing full, adult-level explanations and examples of the more advanced science concepts, principles, and theories that appear in the lessons so that teachers can refresh or enhance their own knowledge of the topics being covered, as necessary.

Selected References

Before It's Too Late: A Report to the Nation from the National Commission on Mathematics and Science Teaching for the 21st Century. Washington, D.C.: U.S. Department of Education, 2000.

Darwin, Charles. *On the Origin of Species by Means of Natural Selection.* Reprinte from the 6th edition. New York: Macmillan, 1927.

DeBoer, George. *A History of Ideas in Science Education.* New York: Teachers College Press, 1991.

Engelmann, Siegfried, and Douglas Carnine. *Theory of Instruction: Principles and Applications.* Eugene, Ore.: ADI Press, 1991.

Gifted Young in Science: Potential Through Performance. Edited by Paul Brandwein and others. Arlington, Va.: National Science Teachers Association, 1989.

Health Framework for California Public Schools, Kindergarten Through Grade Twelve (Revised edition). Sacramento: California Department of Education, in press.

History–Social Science Content Standards for California Public Schools, Kindergarten Through Grade Twelve. Sacramento: California Department of Education, 2000.

Malthus, Thomas P. *An Essay on the Principle of Population.* 1798. Reprint. Amherst, N.Y.: Prometheus Books, 1998.

Mathematics Framework for California Public Schools, Kindergarten Through Grade Twelve (Revised edition). Sacramento: California Department of Education, 2000.

Platt, J. R. "Strong Inference," *Science,* Vol. 146 (1964), 347–53.

Reading/Language Arts/English-Language Development Adoption Report. Sacramento: California Department of Education, 2002.

Reading/Language Arts Framework for California Public Schools, Kindergarten Through Grade Twelve. Sacramento: California Department of Education, 1999.

Science Content Standards for California Public Schools, Kindergarten Through Grade Twelve. Sacramento: California Department of Education, 2000.

Science Safety Handbook for California Public Schools. Sacramento: California Department of Education, 1999.

Appendix
Requirements of the *Education Code*

California's *Education Code* contains a number of provisions that have a bearing on science education in this state. The actual text of the statutes is not included in this appendix because actions of the Legislature and the Governor periodically add to, delete from, or otherwise modify the requirements of law.

Teachers, school administrators, parents (guardians), and others who use this framework are encouraged to become familiar with pertinent provisions of the *Education Code*. For the current text of the statutes, please go to the *California Law* Web site <http://www.leginfo.ca.gov/calaw.html>.

Examples of *Education Code* provisions pertinent to science education (including health and safety concerns) are as follows:

- **Chapter 2 of Part 1 (commencing with Section 200). Educational Equity.** Includes a requirement that teachers endeavor to impress upon the minds of pupils the meaning of equality and human dignity, including kindness toward domestic pets and the humane treatment of animals.

- **Chapter 1 of Part 19 (commencing with Section 32200). School Safety–Public and Private Institutions.** Includes a requirement that eye protective devices be provided for teachers and students under certain conditions and that snakebite kits be supplied for some field trips.

- **Chapter 2.3 of Part 19 (commencing with Section 32255). Pupils' Rights to Refrain from the Harmful or Destructive Use of Animals.** Includes a requirement that pupils who have moral objections to harming or destroying animals be given alternative education projects under certain circumstances.

- **Chapter 6.6 of Part 27 (commencing with Section 49091.10).** *The Edcation Empowerment Act of 1998.* Includes establishment of the right of parents (guardians) to inspect instructional materials and observe instruction.

- **Chapter 2 of Part 28 (commencing with Section 51200). Required Courses of Study.** Includes requirements for the courses of study to be offered and the minimum courses needed for high school graduation.

- **Part 33 (commencing with Section 60000). Instructional Materials and Testing.** Includes establishment of specific controls on state and local adoptions of instructional materials and of penalties for teachers (and other school officials) receiving things of value from publishers.